OIL & VINEGAR

OIL & VINEGAR

More than 1001 natural remedies, home cures, tips, household hints and recipes, with 700 photographs

Bridget Jones

HERMES HOUSE

© 2012 by Anness Publishing Ltd
Illustrations © 2012 by
Anness Publishing Ltd

This 2012 edition is published
for Barnes & Noble, Inc. by
Anness Publishing Ltd

If you like the images in this book and
would like to investigate using them for
publishing, promotions or advertising, please
visit website www.practicalpictures.com for
more information.

ISBN 978-1-4351-4165-0

Manufactured in China

2 4 6 8 10 9 7 5 3 1

PUBLISHER'S NOTE

Although the advice and information in this
book are believed to be accurate and true at
the time of going to press, neither the authors
nor the publisher can accept any legal
responsibility or liability for any errors or
omissions that may have been made nor any
inaccuracies nor for any loss, harm or injury
that comes about from following instructions
or advice in this book. While every effort
has been made to ensure accuracy when
researching this book, the information on the
therapeutic and cosmetic value of vinegar
and oil is anecdotal and is not intended as
a substitute for the advice of a qualified
professional. Any use to which the
recommendations, ideas and techniques are
put is at the reader's sole discretion and risk.
The very young, the elderly, pregnant women
and those in ill-health or with a compromised
immune system are advised against eating
dishes containing raw eggs.

Publisher: Joanna Lorenz

Project Editor: Amy Christian

Photography and styling: Michelle Garrett

Additional photography (recipes): Frank
Adam, Edward Allwright, David Armstrong,
Tim Auty, Steve Baxter, Martin Brigdale,
Nicki Dowey, James Duncan, Gus Filgate,
Will Heap, Amanda Heywood, Ferguson Hill,
Janine Hosegood, David Jordan, David King,
William Lingwood, Patrick McLeavey,
Thomas Odulate, Craig Robertson, Gareth
Sambridge, Sam Smith, Sam Stowell, Jon
Whitaker.

Designer: Sarah Rock

Proofreading Manager: Lindsay Zamponi

Editorial Reader: Lauren Farnsworth

Production Controller: Steve Lang

Recipes: Pepita Aris, Catherine Atkinson,
Mridula Baljekar, Jane Bamforth, Alex Barker,
Gillie Başan, Judy Bastyra, Carla Capalbo,
Georgina Campbell, Miguel de Castro e
Silva, Jacqueline Clark, Roz Denny, Matthew
Drennan, Joanna Farrow, Marina Filippi, Jenni
Fleetwood, Christine France, Silvano Franco,
Yasuko Fukuoka, Brian Glover, Nicola
Grimes, Deh-Ta Hsiung, Christine Ingram,
Becky Johnson, Emi Kazuko, Lucy Knox,
Vilma Laus, Lena Lobanov, Kathy Man,
Maggie Mayhew, Jane Milton, Sallie Morris,
Keith Richmond, Rena Salaman, Jennie
Shapter, Young Jin Song, Marlena Spieler,
Terry Tan, Liz Trigg, Christopher Trotter, Linda
Tubby, Oona Van Den Berg, Sunil Vijajakar,
Stuart Walton, Biddy White Lennon, Jenny
White, Kate Whitman, Carol Wilson,
Elizabeth Wolf-Cohen, Annette Yates.

Additional projects: Stephanie Donaldson,
Jessica Houdret, Renée Tanner.

NOTES

Bracketed terms are intended for American
readers.

For all recipes, quantities are given in both
metric and imperial measures and, where
appropriate, in standard cups and spoons.
Follow one set of measures, but not a
mixture, because they are not
interchangeable.
Standard spoon and cup measures are level.
1 tsp = 5ml, 1 tbsp = 15ml, 1 cup =
250ml/8fl oz.

Australian standard tablespoons are 20ml.
Australian readers should use 3 tsp in place
of 1 tbsp for measuring small quantities.

American pints are 16fl oz/2 cups. American
readers should use 20fl oz/2.5 cups in place
of 1 pint when measuring liquids.

Electric oven temperatures in this book are
for conventional ovens. When using a fan
oven, the temperature will probably need to
be reduced by about 10–20°C/20–40°F.
Since ovens vary, you should check with your
manufacturer's instruction book for guidance.

The nutritional analysis given for each
recipe is calculated per portion (i.e. serving
or item), unless otherwise stated. If the
recipe gives a range, such as Serves 4–6,
then the nutritional analysis will be for
the smaller portion size, i.e. 6 servings.
The analysis does not include optional
ingredients, such as salt added to taste.

Medium (US large) eggs are used unless
otherwise stated.

CONTENTS

INTRODUCTION

We are all familiar with the basic culinary uses of vinegar and oil, but this book sets out to explore the great all-round potential of these everyday ingredients.

The far-reaching benefits of these ancient ingredients have been recognized in many different cultures throughout the ages. Although nowadays they are most often put to use in the kitchen, vinegar and oil can be used for a huge variety of tasks around the house. Together or separately, these hardworking kitchen staples have traditionally fulfilled many roles beyond their food and household uses, including medicinal purposes, as beauty aids and in social and religious rituals. As two of the oldest manufactured ingredients, these two substances have evolved and progressed to develop a harmonious partnership.

Astringent vinegar

Together, the two substances are often referred to as 'oil and vinegar', with oil coming first; however, vinegar is more likely to precede oil in practical use, as the acid, solvent and cleanser. Vinegar is the aggressor of the pair, an astringent or type of antiseptic, used to clean and prepare foods and remove impurities, and as a cleaner in the home. Its acidic

Above: Oil has been used in cooking since ancient times. It is now found all over the world in many different dishes.

characteristic makes vinegar the primary long-term food preservative, used to prevent the growth of bacteria and deterioration. The sharpness of vinegar has the effect of spotlighting some flavours, accentuating them and bringing out distinct contrasts.

Soothing oil

Oil, on the other hand, is the soother of the pair. In every aspect of its use, oil is used to enrich whatever it comes in contact with. While it does bring flavours, textures and colours to life, it also creates harmony. This is easy to see in culinary use, where oil enhances dishes or is used as a rich cooking medium. When oil is combined with a main food and other ingredients, it draws the flavours together and has a mellowing effect. It gently highlights some tastes or colours in cooking while subduing others.

In household use, oil follows vinegar after the vigorous cleaning is over. When wood is cleaned with vinegar, the dirt and residue of wax and other polishes is removed before oil-based finishes can be applied to bring out the colour and unite the visual features of the wood. The same is true of some metals, for which vinegar may be used as a cleaner while oil will act as a polish and protective layer to help prevent corrosion.

Culinary essentials

Steeped in history but still vital today, vinegar and oil have survived advances in food production as well as fashion, not only retaining their essential places in the household but also increasing in relevance as we learn more about their beneficial properties. It is difficult to imagine how we would live without vinegar and oil. Both are used widely in the preservation of ingredients to make everyday food items, whether in condiments, sauces, salsas, relishes, bottled or canned salads (antipasti or tapas style) and dressings.

Most people have a basic choice of vinegar and oil in the kitchen, but this book aims to show that a wider selection of even just a few different vinegars and oils can be a real boost

Right: Vinegar has long been an essential culinary ingredient, and is used in the preservation of many foods.

to the dining experience. Simply drizzling vinegar and oil over dishes or ingredients before or after cooking will bring flavours together, accentuating some, making others more mellow and emphasizing differences. These two ingredients are able to perform culinary magic!

Utility value

Being described as utility substances does not sound very glamorous, yet vinegar and oil are highly regarded as cleaning materials. Instead of harsh chemicals and the unwanted films and residues from cans of spray-and-wipe solutions, these natural substances are ideal for restoring fine furniture and for helping to keep the home clean, fresh and eco-friendly.

In the garden vinegar and oil can provide quick, easy and environmentally friendly solutions for many simple problems. From pest control to wood preservation, vinegar and oil are increasingly appreciated, and as natural, non-toxic substances, they are safe to use around children or pets.

Health and beauty

Vinegar and oil have long-standing reputations for promoting good health and enhancing beauty. Yet again, the positive partnership between astringent, antiseptic and cleansing vinegar and soothing, restorative oil is important. Including them as part of a balanced diet will contribute to overall well-being, while externally both vinegar and oil are known to perform well in first aid and also provide long-term benefits to the health of skin, hair and nails.

Manufacturers of sophisticated beauty products often emphasize the natural simplicity of their core ingredients while making incredible claims about their potential. There is a growing realization however of the benefits of using natural ingredients, such as vinegar and oil, as part of a beauty regime. Simple preparations can be made at home with store-cupboard items for a fraction of the price of expensive, branded products.

Previous generations have made the most of the amazing properties of vinegar and oil, along with other natural ingredients, to complement commercial beauty products and medicines. With a growing awareness of the need to respect our environment and an increased emphasis on the importance of including simple unprocessed products in our health and beauty regimes, this is more relevant today than ever before.

Left: Beyond the traditional culinary uses of vinegar and oil, there are many more exciting applications waiting to be discovered.

Above: There are many uses for oil outside the kitchen. It is particularly good for polishing and finishing wooden surfaces.

Inspiring selection

From Italian balsamic vinegar, British malt vinegar and Scandinavian rapeseed (canola) oil, to Moroccan argan oil and West African palm oil, the huge variety of vinegars and oils that are available today can in themselves be a source of inspiration for experimentation. A bottle of coconut vinegar, for example, cries out to be used in more exciting ways than as a simple dressing.

The full range of vinegars and oils can also be explored for use in health remedies. Natural organic cider vinegar is traditionally prized for its positive health benefits, but there are also other fruit and grain vinegars, as well as oils extracted from a wide variety of seeds and nuts. These are readily available from supermarkets, pharmacies, healthfood shops and online.

Exciting potential

The chapters in this book offer a glimpse of the history of two ingredients that we have used for thousands of years. The ancient origins of vinegar and oil are discussed, showing the fascinating ways that they have been put to use over the centuries. Detailed directories outline the many different types of vinegar and oil, before practical chapters list the various ways that they can be used for health and healing, for natural beauty, around the home, in the garden and of course in the kitchen.

It is not so long since these simple substances were put to extensive use in every corner of every household. Rediscovering the many uses for vinegar and oil is exciting, and trying some of the remedies, potions and lotions that our grandparents used, or remember, is great fun, with often surprising and dramatic results.

VINEGAR

From accidental discovery in ancient
Greece to the sophisticated fermentation
methods of modern vinegar factories, vinegar
has a haphazard but fascinating history.
Today there is global appreciation of the
remarkable properties of this sharp substance
that helps with healing, assists in protecting
against illness and has the ability
to magnify the flavour of food.

Above: Airtight bottles of vinegar should be stored in a cool, dark cupboard.
Left: Home-made flavoured vinegars make attractive and interesting culinary gifts.

VINEGAR IN HISTORY

Vinegar is an ancient and global ingredient that has been utilized by human beings for so long that its exact origins, or discovery, cannot be traced precisely.

Vinegar is an acid produced when certain bacteria attack alcohol. The first experiences of vinegar were unwelcome accidents of nature. In ancient times, when wine turned sour, the resulting liquor, an unwanted by-product of wine making, was seen as a disaster rather than a triumph.

The useful properties of the soured wine, or vinegar, as a preservative were probably first discovered by chance some 10,000 years ago. It was later, around 5000BC, that the Babylonians began to produce vinegar intentionally and put it to practical use both as a preservative and for culinary purposes. They discovered that the soured wine could be flavoured for use as a condiment. From their wines made from date palm, they produced vinegar infused with herbs and spices.

Egyptian archaeologists have unearthed vessels from around 3000BC believed to have stored vinegar.

Above: Hippocrates prescribed vinegar and honey to his patients to balance the 'humours' in the body.

Evidence for Greek and Roman use of vinegar as a condiment includes the remains of bowls, called oxybaphon and acetabulum, that were used for serving vinegar at the table so that bread could be dipped into it. The early use of vinegar for moistening bread continued through Biblical times.

Vinegar and the Bible

There are many Biblical references to vinegar, but the most famous must be to a sponge soaked in wine vinegar offered to Jesus as he was dying on the cross: 'When Jesus therefore had received the vinegar, he said, "It is finished." He bowed his head, and gave up his spirit.' (John 19:30) Generally interpreted as an act of mockery, other mentions of vinegar indicated that it may have been an act of kindness. Vinegar was believed to draw out every last bit of moisture from the skin in the mouth to provide some brief comfort in swallowing, and it was also believed to have restorative qualities.

Among earlier Old Testament references to wine vinegar as a drink, there was an instruction to Moses to forbid the Israelites from drinking vinegar made from wine (as well as any other fermented drink). For her work harvesting barley, Ruth was offered bread to dip in wine vinegar, a sign of friendship and acceptance.

One of the Proverbs of Solomon that warned against employing the lazy (sluggards) indicated that vinegar was probably widely consumed and that drinking too much was known

Above: An angel holds a vinegar-soaked sponge, representing the vinegar offered to Jesus on the cross.

not to be good for the teeth. Another proverb implied that the reaction of vinegar mixed with nitre, a name for sodium carbonate or washing soda, was widely understood.

Hippocrates and health

Before Biblical references to vinegar as a restorative drink, it had been widely regarded as a valuable aid to health by the Egyptians, Greeks and Romans. The health benefits of vinegar were acknowledged by Hippocrates, the Greek physician (460–370BC). Often referred to as the father of modern medicine, Hippocrates documented the importance of hygiene and based his theories on the need to achieve a balance for good health. The balance he referred to was with reference to

Above: Hannibal used vinegar to break down rocks as he crossed the Alps.

the 'humours' of the body, thought to be blood, phlegm, yellow and black bile. He prescribed both cider vinegar and honey, together and separately, as a balancing tonic to help cure many ailments.

During the14th to 18th centuries when the plague was evident throughout Europe, vinegar was used as an antiseptic and rubbed into the skin in an effort to avoid infection. The disease was so virulent in 18th-century France that prisoners, who were considered to be expendable, were released to bury the dead. A group of thieves who stole from the corpses they buried survived by cleaning themselves with vinegar to avoid infection and by drinking vinegar infused with garlic, a natural antiseptic. Garlic vinegar is still sometimes referred to as 'four thieves vinegar' for its legendary protective powers against the plague.

Vinegar was used as a deodorizer as well as an antiseptic. When the gentry and ladies of the 17th and 18th centuries walked out on streets that reeked of sewage and waste, they masked their mouths and noses with sponges soaked in vinegar to protect themselves against both the odours of the common people and the potential for contamination.

Small silver boxes and compartments in the tops of walking sticks were designed to hold the little vinegar-soaked sponges.

Military use of vinegar

From ancient to modern times, and worldwide, vinegar was used to promote good health among military ranks. Roman soldiers consumed vinegar to promote good health. They combined vinegar with honey and diluted it to make drinks.

Around 218BC the Carthaginian general Hannibal's army was said to have used vinegar to assist in breaking up rock when clearing a way through the Alps in the advance on Italy. Fires were lit to heat impenetrable rocks, which were then soaked with vinegar. The heating and soaking in acid softened the rock enough to allow the soldiers to break them down into boulders that could be cleared away.

In the 17th century, Louis XIII of France was said to have paid handsomely for vinegar to cool and clean the cannons used in battle. Cleaning the metal with vinegar helped to prevent rusting as well as removing dirt.

During the American Civil War (1861–5) apple cider vinegar was consumed by soldiers to treat scurvy, and as late as the beginning of the 20th century, across the Atlantic, soldiers in the First World War resorted to using vinegar as an antiseptic for those wounded in battle.

Clever Cleopatra

Hannibal was not the only leader to use the powers of vinegar for dissolving solid objects. The Egyptian queen Cleopatra (68–30BC) used vinegar for a more frivolous purpose, to dissolve a pearl. She drank the resulting liquid as a clever way of winning a bet about consuming the most expensive meal by literally consuming the huge cost of the pearl.

Above: Cleopatra won a bet with Mark Antony by dissolving a pearl in wine vinegar.

THE ENDURANCE OF VINEGAR

Production methods may have progressed, and many varieties of vinegar developed, but the basic uses for vinegar have remained the same through the ages.

Vinegar is a product with a well-founded history as an essential ingredient and basic household item. Different vinegars have maintained their roles for thousands of years, whether for cleaning, health, beauty, cooking or preservation. Vinegar is a product which has to be made, rather than occurring naturally, but efforts put into producing vinegar have long been rewarded by its usefulness.

European cottage industries

The ancient Greeks, Romans and Egyptians set up the first formal vinegar production and storage systems, but following the decline of the Roman Empire, wine (and therefore vinegar) production decreased. Wine and beer were produced on a small scale and so when the value of vinegar as a culinary and household treasure was rediscovered in medieval Europe, its production was very much a cottage industry. Vinegar was a by-product of the alcohol production process, and was made in home breweries and small vineyards, rather than in large specialist vinegar factories.

In Paris, vinegar made from the produce of local vineyards was sold by street vendors, who would roll barrels or carts from door to door. The vendors hailed housekeepers and persuaded them of the quality of their 'vinaigre'. It was during the 14th century that a commercial vinegar production corporation was set up in Paris.

In Great Britain the production of malt vinegar, or 'alegar', was so profitable that a vinegar tax was established in 1673 by an Act of Parliament.

Above: This 17th-century engraving shows a French vinegar seller.

Development of Italian vinegars

The town of Modena in Italy had also been producing and maturing vinegar since the Middle Ages. Records indicate that the sweet vinegar syrup of Modena was sold from the 14th century. By the 17th century fashionable households boasted their own stores of 'balsamico', a product that had developed an amazing reputation. It was even said to be capable of bringing the dead back to life. Modena remains the centre for balsamic vinegar production today (along with nearby Reggio Emilia). The name 'Aceto Balsamico Tradizionale di Modena' is protected by the European Union's Protected Designation of Origin (PDO).

Cider vinegar in the United States

Introduced by early European settlers, apples were used in the United States for eating, cooking, and to feed livestock.

Above: Cider vinegar has been the most popular vinegar in the US since the 18th century.

Apples were also pressed to make apple juice, a portion of which was fermented into cider, allowing it to be stored over the winter months. Some of the cider was allowed to ferment further into cider vinegar, which was prized for its many uses – as a medicine, a condiment, a preserving liquid, for cleaning and a host of other household duties.

During the 18th and 19th centuries, American farm labourers would consume 'switchel', a drink made from cider vinegar mixed with water and ginger and sweetened with honey. Switchel served as an early energy drink for the exhausted workers out in the field. By the early 1800s cider vinegar was selling for three times the price of apple cider.

Chinese and Japanese vinegars

Vinegar has always featured in oriental medicines and diets. Although much of the history of the discovery of vinegar and development of commercial production focuses on the wine and malt vinegar industries, rice has been used for producing vinegar in both Japan and China since ancient times (Chinese records dating back to 1200BC include references to vinegar). Rice vinegar is still a vital cooking ingredient in these countries and is used in Chinese medicine.

Value of vinegar now

The fundamental uses of vinegar have stayed much the same for several hundred years, with increased interest in the role that vinegar can play to promote health and well-being. Even though there are now many expensive vinegars available, standard vinegar still has a high profile in the supermarket and a place in home store cupboards.

Everyday chores As the array of harsh household cleaners and chemicals has expanded, it is good to find a simple

Above: Vinegars made from different ingredients are found all over the world. Vinegar made from rice was used in ancient Chinese medicine as well as in cookery.

yet effective alternative that is easy to use and inexpensive. Vinegar is a practical choice for use as an everyday home-cleaning product as well as for renovating and deep cleaning.

Natural balance At the same time as trying to minimize household use of potentially harmful products, many people are also looking for natural ways of building up resistance to illness and infection, and of averting or counteracting relatively minor conditions. Instead of dashing to the pharmacy for every little complaint, there is a growing trend to use simple cures for minor irritations.

Simple skin care Legend has it that Helen of Troy (described as the most beautiful woman ever in Greek mythology) bathed in vinegar. While this may be a little extreme, vinegar has been used for centuries for refreshing baths, face and hair washes and it is still popular. Alternatives are increasingly being sought to replace high-tech beauty products. Simple, inexpensive blends are welcomed by those who would prefer to minimize their use of complicated and often expensive beauty ranges.

Gourmet ingredient The history of vinegar suggests that it was originally most valued for its health-giving properties. It was used in even greater quantities than it is now as a restoring and refreshing drink, and to aid digestion.

Vinegar has been used in preservation throughout history, for pickling fish and meat as well as fruit and vegetables. It plays a vital role in the preparation of classic French sauces and in the sweet-sour or hot-sour flavours of Chinese cooking.

An increase in the availability of different types of vinegar has led to a growing appreciation of the sometimes subtle distinctions in flavour between the many varieties. In recent decades, vinegar has become more than a tonic, vital culinary acid or simple souring agent, and has become increasingly seen as a gourmet ingredient. There are many that are used with discretion for their flavour nuances. Vinegar has developed from a basic store-cupboard ingredient to a highly-prized condiment, of which there are now many types to choose from.

FROM ALCOHOL TO VINEGAR

In essence, vinegar is the product of a natural process of fermentation, for example, of fruit, barley or rice, followed by bacterial activity.

The basic ingredients of vinegar, such as grapes, other fruit, malted barley or rice, are fermented to produce alcohol and then a bacteria culture is added to produce the acidic vinegar. The flavour of the original ingredients is often retained and, being acidic, the vinegar tastes sharp. The combination of sharpness and flavour nuances gives vinegar its culinary value.

Fermentation

The first step towards vinegar being made is a process called alcoholic fermentation. The process happens in nature, for example when yeasts react with the natural sugar (glucose) and water in ripening fruit to produce alcohol in the form of ethanol.

Above: Vinegar is formed when bacteria in fermenting alcohol produce acetic acid.

To complete the chemical equation, carbon dioxide (CO_2) and heat are given off as by-products of the reaction. Compounds called esters also develop during the reaction, producing fruit flavours in the vinegar.

The commercial fermentation process is a controlled version of what happens in nature. Wild yeasts do not necessarily produce pleasing flavours, so specific strains of yeast are used for making wines and beers. By controlling the yeasts and the process, the fruit base (or grain in the case of beer) is fermented under controlled conditions to produce alcohol with a good flavour. When all the sugar has been used the process will stop, because the yeast needs food in order to live and multiply. Alternatively, when the alcohol level becomes too high, the yeasts will not survive.

From alcohol to acid

Anyone who has tried wine-making or brewing their own beer at home will know that the sugar concentration and temperature, along with cleanliness, are very important. Equipment has to be sterilized and air must be excluded from the fermenting liquid for any chance of success, otherwise the wine or beer can develop very unpleasant off-flavours. In the same way that yeasts are present in nature, some bacteria naturally present in fruit or carried in the air will work on the alcohol and break it down into acid.

MOTHER OF VINEGAR

It is vital to preserve the particular strains of aerobic bacteria, or *acetobacter aceti*, which are used to ferment the wine or other base and which produce acetic acid along with good flavour.

In the traditional production methods, the bacteria form a slimy, slightly cloudy string or surface on the fermenting vinegar and this is known as the mother of vinegar, vinegar mother, or simply the mother. The mother of vinegar forms on the surface because there it has access to air.

In classic techniques, enough mother of vinegar is retained to continue the vinegar-making process. Before pasteurization was a standard process used for inexpensive vinegar, a half-empty bottle of vinegar stored in a warm place, or in direct sunlight, would soon develop a mother of vinegar which could clearly be seen floating in the liquid.

The bacteria need a supply of oxygen to produce acetic acid from the ethanol, so there has to be air in the brewing container. Fermenting beer or wine left open to air, perhaps in equipment that was not properly cleaned after previous use, is likely to become acidic and more like vinegar, which is a disaster for the home wine maker or brewer who anticipated sipping a successful liquor but found that they were left with sharp-tasting vinegar instead.

From acid to vinegar

In commercial vinegar production, the bacteria *acetobacter aceti* are desirable, and are deliberately added to the fermented alcohol. When there is oxygen present, the bacteria will turn the alcohol into acetic acid, which is one of the main components of vinegar.

Many centuries ago the first vinegar was made by the accidental presence of bacteria in fermenting alcohol. Once vinegar was valued in its own right, the process was encouraged and vinegar was eventually put to positive use as a preservative. Over the generations, vinegar has grown from being a by-product of wine, beer or other alcohol, and a large specialist vinegar industry has developed.

Above: If a bottle of unpasteurized vinegar is left opened, the presence of oxygen means that a mother will soon develop.

Vinegar producers take a great deal of care to select beer or wine of the best quality for their vinegar and to ensure that the conditions are exactly right for avoiding contamination by any unwanted bacteria that might produce any undesirable off-flavours in the vinegar.

ACETIC ACID

Acetic acid, which is also known as ethanoic acid, is the main ingredient (besides water) in vinegar, and gives it its sour taste. It is important to differentiate however between vinegar made from a traditional ingredient base and diluted acetic acid which is made from ethanol and has none of the positive qualities of vinegar.

Vinegar is made from a base ingredient such as barley, fruit or rice. The naturally occurring yeasts in these ingredients react with sugar and produce compounds called esters that contribute the flavour to the vinegar. Instead of a food base, industrial acetic acid can be produced from ethanol (pure alcohol).

Acetic acid is very sharp and often used in preserves and pickles instead of vinegar. It may be useful as a preservative but it contributes no flavour and makes the food taste particularly bitter; even sweetened pickles made with acetic acid have an unmistakable harshness. The best advice is always to carefully check the labels on pickles and other products that would traditionally use vinegar (such as salad dressings and condiments) because acetic acid may be used instead. It is also a good idea to check out the labels on wine vinegars, particularly the less expensive types that look as though they are full flavoured and matured — they may actually be simple vinegars combined with sweetening and flavouring ingredients.

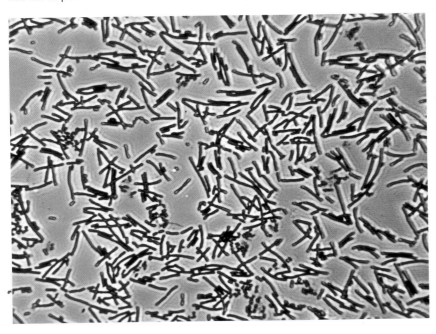

Above: The acetobacter aceti *bacteria which produce acetic acid from alcohol.*

VINEGAR AS AN ACID

Vinegar is an acidic liquid resulting from the fermentation of alcohol. Its acidity is an important factor for its many household uses.

Knowing a little bit about acids and how they relate to, and react with other substances helps to explain why vinegar is so useful for certain household purposes as well as in cooking.

About acids and alkalis

In chemical terms, pure water is neutral, neither acid nor alkali. Acids and alkalis, known as bases, are measured on a scale measured by pH, which stands for potential hydrogen or power of hydrogen and is related to the concentration of hydrogen ions in the liquid. The scale ranges from 0 to 14, with pure water in the middle with a pH of 7. Acids have a pH below 7, getting more acidic as the pH is lowered, so a strong acid would have a pH of 2 to 0.

Above: Wine vinegar is created when fermented grapes become acidic.

Alkalis or bases have a pH above 7, with a pH of 12 to 14 being very strong. The scale 0 to 7 is not measured in straightforward increments of 1 but in logarithmic stages. A difference in pH of 1 represents a tenfold difference in strength. Although the difference in figures may seem small, the actual difference in strength of the liquids will be large. A meter is used to test pH.

Properties of acids

Acids can be solids as well as liquids, for example, citric acid is available as crystals as well as in liquid form. Tartaric acid is found in fruit, especially grapes, but a related solid form (or acid salt) of potassium hydrogen tartrate is commonly known as cream of tartar.

Acids have a sharp or sour flavour – citric acid makes citrus fruit taste sharp; the acetic acid in vinegar makes it taste sour. Lactic acid makes some dairy products taste the way they do, for example sour milk, cream or tart soft cheeses have a slightly sharp, tangy or fresh flavour.

Reactions between acids and alkalis

When acids and some alkalis or bases are combined, they react and are neutralized. In the process, another chemical compound or salt is produced and gas given off. This type of chemical reaction can be useful. For example, in baking, mixing cream of tartar (an acid) with bicarbonate of soda (an alkali), adding water and then

TYPICAL pH VALUES

ACID	pH 0	strong hydrochloric acid
	pH 1	battery acid
	pH 2	lemon juice, gastric acid, acetic acid
	pH 3	vinegar, carbonated soft drinks, orange juice, cream of tartar
	pH 4	citric acid, tomato juice, skin, hydrogen peroxide
	pH 5	coffee, handwash liquid
	pH 6	urine, saliva
NEUTRAL	pH 7	distilled water, milk, blood, semen
	pH 8	egg white, bile
	pH 9	bicarbonate of soda (baking soda), borax
	pH 10	washing powder, milk of magnesia
	pH 11	bleach, laundry ammonia
	pH 12	dishwasher powder, photo developer
	pH 13	lime, lye
ALKALI	pH 14	caustic soda

ACIDS

Vinegar is a mild acetic acid, which is not usually damaging to skin, but strong acids will burn the skin. Strong acids will also attack fabrics (natural and manmade) and cause them to disintegrate.

heating produces a gas, carbon dioxide, that makes the mixture rise. Bubbles of the carbon dioxide are trapped when the mixture bakes. Vinegar, another acid, is used as a raising agent in some baking recipes, when mixed with bicarbonate of soda in the same way. The reaction between acid and alkali is also used in traditional effervescent tablets available for relieving indigestion.

Acids and metals

The reaction between acids and metals can be unwelcome when using cooking equipment, especially uncoated copper, iron or aluminium pans. The flavour of the food can be tainted by the metal and the

Above: Citrus fruits such as oranges, lemons and limes contain citric acid.

reaction may also produce unwanted toxins. This reaction, although usually seen as a negative one, is however sometimes deliberately used in food preparation. Some old-fashioned cooking equipment was uncoated and this was sometimes valued for the effect it had on the food.

Above: Lactic acid gives some dairy products a sharp, tangy flavour.

For example, when copper preserving pans were used for making jellies and sweet preserves the acid from the fruit reacted with the uncoated copper giving the preserve (if properly prepared) a particularly sparkling clarity, albeit with a small unwanted dose of metal.

Generally, however, the reaction between vinegar (an acid) and metal should be avoided in cooking. When making chutneys and pickles with vinegar as the preservative, care must be taken. When bottling the preserve, uncoated or damaged metal lids should not be used to seal jars or bottles, as the acidic preserve will rapidly corrode the lid, causing damage and ruining the preserve.

The same effect can be found when using aluminium foil as a direct wrapping on food, for example around a rich fruit cake. The acid from the fruit will rapidly corrode the foil, reducing it to a powder. Wrapping the item in baking parchment first prevents the contact and therefore the reaction.

Above: The acidic properties in vinegar make it an excellent base for marinades, used alone or with other ingredients.

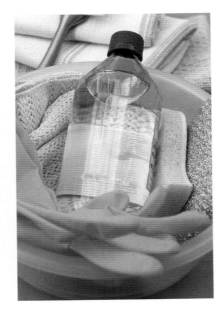

Above: The acetic acid in vinegar allows it to cut through grease and dirt when used as a cleaning agent around the home.

COMMERCIAL VINEGAR PRODUCTION

In vinegar factories, a carefully controlled process guarantees that the fermentation, environment and maturing process produce perfect vinegar.

Methods of production for vinegar ensure that the correct bacteria are used, that the fermentation base is of suitable quality, and the environment for souring allows the required bacteria to work. It is important that foreign fermentations are prevented – these may ruin the flavour of the vinegar. Temperature and exposure to air are standardized along with the length of fermentation. When one batch of vinegar is produced, a small percentage, containing the 'mother', is retained and used to begin the process again. Today there are several different methods used for commercial vinegar production.

The Orléans method

France is famous for refining vinegar production and developing the Orléans method, also known as the continuous method. Orléans, an ancient city of pre-Roman existence on the Loire, was liberated from British rule by Joan of Arc in 1429, by which time it was already famous as a centre for food and wine collection and distribution, as well as for vinegar production. The vinegar industry grew because wines brought to Orléans would often arrive soured in the cask.

King Charles VI officially recognized the vinegar merchants' corporation in 1394. Some 200 years later, just as there were *confréries* or associations, orders or brotherhoods to protect local wine-making standards and traditions, the local profession for vinegar merchants was recognized by King Henri III in 1580.

Local distributors had developed the trade by taking the *vin aigre*, or sour wine, and fermenting it to produce good-quality vinegar rather than letting it go to waste. The process was used only for good-quality wines, which were slowly fermented in the barrel under controlled temperature conditions. Local barrel makers set the standards, ensuring that the barrels were aerated and not over-filled so that the true flavours of the original wine were preserved and carried into the vinegar. Once it had been fermented, the vinegar was also matured.

Vinegar production developed and by the 18th century, Orléans was established as the centre for wine vinegar. The slow and careful process is still used today by specialist vinegar producers and the Orléans method is recognized as a sign of great quality.

Above: In the Vinegar Factory in Hamburg, *Gotthardt Johann Kuehl, 1891.*

Above: Wine vinegars are matured in oak barrels for at least six months.

Above: Balsamic vinegar matures for many years then is tested by skilled producers.

Generator method

The Orléans method of vinegar production is time-consuming. A new generator method, which speeds up the process by fermenting bacteria over a bed of wood shavings, was developed early in the 19th century by a German vinegar producer, Schuezenbach. It is also known as the German or Schuezenbach method as well as the 'quick vinegar' or 'fast acetic' method.

A large surface area of wood shavings (originally beech) was used as the base for the vinegar culture, providing greater exposure to oxygen for the vinegar mother, which reduced the time needed to produce the acetic acid. The liquor was sprayed evenly over the wood shavings. To achieve the required level of acidity, the process could be repeated. Alternatively, the liquor was passed through several aerated barrels or generators over a few days. This was not as efficient as spraying the liquor.

In traditional malt vinegar production, the liquor was trickled down threads on to a bed of beechwood shavings or charcoal that formed the fermentation base. Base ingredients such as beer, brandy, whisky, molasses and honey were used in this way for fermentation and acetic acid production. The disadvantage of using this quicker method was increased evaporation of the alcohol and therefore loss of acetic acid. The quick fermentation generated sufficient heat to destroy some of the natural flavours that would be retained by a cooler, slower process.

Instead of maturing this new vinegar which had been made by the generator or 'quick vinegar' methods, flavouring ingredients, such as caramel (which gave a rich brown colour) were added before the vinegar was sold.

Submerged fermentation method

Further developments soon made vinegar production even quicker. The vinegar mother was submerged in the liquid and air pumped through it for continuous aeration. Temperature and exposure to air were highly controlled and the whole process was sped up to produce a product of reliable basic quality. The disadvantage was that the character or intense flavour found in slow-fermented wine vinegars was lost. This method is used today and produces vinegars for commercial pickling and sauce production.

Finishing fermentation

Once the vinegar has fermented, it is either filtered to remove the mother and debris or may be cleared in a similar way to wine. To prevent unwanted further fermentation, air has to be excluded from the vinegar. Some types of vinegar are pasteurized to kill off the remaining bacteria.

Maturing wine vinegar

The best wine vinegars are matured in barrels that have been used for the purpose for many generations. Skilled vinegar makers ensure that the conditions are right to promote maturation and they assess the quality of their product, checking progress for the point when the wine vinegar has developed yet not lost its fresh acetic quality. While some wine vinegars may need a minimum of six months ageing, the balsamic vinegars of Italy in particular are known for long maturation of up to 50 years or more.

HOME-MADE VINEGAR

Making your own vinegar is fascinating and can be fun, but there are a few pitfalls to avoid and tips to follow for reliable results. Use live – unpasteurized – vinegar as a starting point.

The idea of making vinegar may seem novel now but in the past it was fairly standard practice to keep a vinegar mother on the go for souring wine. Before wine vinegar was readily available, adding the remnants of a bottle of wine to vinegar to ferment down was a good way of making better quality vinegar. It is certainly worth an attempt at making some vinegar – if a decent mother develops, then more adventurous experiments with home-made fruit wines and better-quality vinegar or sherry may be worthwhile.

Traditional methods

The tried and trusted way to make vinegar is to use a barrel with a tap in the bottom and pour in live, organic wine, then add some live (unpasteurized) vinegar that will ferment. The fermentation takes place on the surface of the vinegar, forming a scum of bacteria, the mother. The bacteria need air, so the barrel must not be filled, and it needs to be kept in a warm place. The vinegar can be drawn off through the tap and more wine added. The larger the quantity, the longer the souring will take.

Basic vinegar-making

A simple way of experimenting is to use some wine or cider and live vinegar. Dilute strong wine with water and mix it with some live vinegar. Use a glass or glazed container, not metal or lead crystal – a large jug (pitcher) or crock is ideal for an initial experiment. Cover the jug with muslin (cheesecloth)

Above: A container with a tap in the bottom is useful when making vinegar.

to allow air in but keep dust and dirt out. Then leave in a warm place, preferably somewhere dark and out of direct sunlight. The temperature should be constant (as for home brewing or wine-making), as off-flavours develop if the fermenting liquid fluctuates between very hot and cold conditions.

The mother of vinegar should form on the surface and this must not be disturbed. Use a drinking straw to remove small samples of the souring wine from the bottom of the jug and check progress over 2–4 weeks. The wine should develop a vinegar-like aroma.

Once the vinegar has reached a strength you are happy with, or if it begins to lose its flavour, it is ready. Once fermented, carefully syphon off

MAKING VINEGAR

1 To make your own vinegar at home, mix strong wine with water and live vinegar.

2 Cover with muslin (cheesecloth) and allow the mixture to ferment in a warm, dark place.

Above: When making vinegar the choice of base ingredient – for example, cider, wine or beer – is important. For a truly do-it-yourself approach, use home-made alcohol.

the vinegar into a separate container, leaving the mother behind with a small amount of vinegar for making the next batch. Strain, place the vinegar in a clean airtight container and store in a cool place to clear. The vinegar should be allowed to mature for at least six months so that the flavour develops. To prevent further unwanted fermentation, Campden tablets (potassium or sodium metabisulphite – available from a brewing supplier) can be added to kill off the bacteria before the vinegar is matured.

Serious home-vinegar production

To produce reliable, good-quality vinegar at home means being more stringent about methods, sterilizing equipment to keep unwanted bacteria from the fermentation and checking both alcohol levels in the wine and acid levels in the vinegar. Home brewing and wine-making suppliers are a good source of equipment, from glazed fermentation crocks and heated mats or jackets designed to maintain constant temperatures for fermentation, to equipment for checking alcohol levels before starting and acid-testing kits. Some suppliers may be able to provide detailed information on making vinegar at home, including supplying suitable starter bacteria. Of course the choice of cider, beer or wine is also important and, with experience, this can be home-made.

TIPS FOR MAKING VINEGAR

- Check out suppliers of the bacteria culture (acetobacter), for example on the Internet or through local wine-making suppliers. Remember, though, that vinegar is anathema to wine makers and some may be affronted by an enquiry about acetobacter; however, many enthusiastic home brewers have experimented with vinegar and are very helpful. If acetobacter is not available, use live organic vinegar as a starter.

- Use organic cider, beer or wine. Alcohol that is chemically treated to prevent further fermentation will not be suitable. For example, potassium sorbate and/or Campden tablets or the equivalent are added to stop micro-organism activity and this will prevent a vinegar mother from forming.

- Check the alcohol content. It should be 5–7 per cent, therefore stronger wines should be diluted with water.

- Acid testing kits (from wine-making suppliers) should be used for checking the acidity level, which should be at least 5 per cent.

- Find somewhere clean and constant in temperature for keeping the culture working. The temperature should be about 16–27°C/60–80°F. Spells of warm weather are ideal, but make sure the temperature does not rise too high during the day or drop sharply at night.

- Sterilize equipment using a solution recommended for baby feeding equipment, rinsing everything thoroughly. Wash the cloth for covering the container in sterilizing fluid.

- The mixture should begin to smell sharp within a few days and take up to a month to become vinegar.

- Strain the vinegar through coffee filter paper when it is sufficiently acidic.

- If the result is good, save some of the vinegar and mother to make another batch.

DIRECTORY OF VINEGARS

From crystal-clear distilled malt vinegar to glossy, dark treacle-like balsamic, there is a vinegar to suit each food, every dish and all occasions. Beyond the kitchen, there is a choice of basic vinegars for household use and several types are ideal for simple beauty preparations. All vinegars have some positive contribution to make to health, from antiseptic qualities to promoting good digestion. This section provides a guide to the appearance and properties of all the key types of vinegar.

Left: Vinegars vary in colour and clarity according to the ingredients from which they are made.

A VARIETY OF VINEGARS

Vinegar is an ancient and global ingredient, so it is not surprising that there is now an amazing selection from which to choose. Different vinegars suit different purposes, so choose with care.

The first discovery of vinegar was in soured wine, but since then many different varieties of vinegar have been manufactured. Ale was also traditionally used for producing 'alegar' or sour ale. Beer, ale, cider, mead, sherry or perry (fermented from pears) can all be used as a base for vinegar, or fruit juices or mashes may be the starting point for fermenting through alcohol to acetic acid. Besides grapes or apples, raspberries and other berries, or dates, figs and stone (pit) fruit are traditionally used to make vinegar. Dried fruit, as well as fresh, can be fermented into vinegar. Sugar cane, molasses and glucose or corn syrup are also used.

Rice is the base for many Chinese and Japanese vinegars but other grains, such as millet and barley, can be used. Cherries, peaches, dates, grapes and honey have all played a part in Chinese vinegar-making history. Other South-east Asian vinegars are made from coconut water and palm sap.

One of the most unusual base ingredients must be whey, the liquid left after producing cheese curds, which was fermented by a Swiss vinegar producer early in the 20th century. It is unlikely to feature on the average supermarket shelf, but whey vinegar is still available as a healthfood product.

Choosing vinegars

The type of vinegar should not be seen as a guide to quality but of taste and style, and suitability for different uses. Quality depends on the basic ingredients used, the fermentation process and whether or not the vinegar has been matured. Select vinegars to suit different culinary purposes or other needs; for example, a malt vinegar may be suitable for certain pickles while wine vinegar may be ideal for simple dressings to coat everyday mixed salads. Superior wine vinegar may be reserved for dressing individual fruit or vegetables, to deglaze cooking pans or for adding piquancy to sweet dishes.

The best advice is to experiment with different types and brands, buying small quantities that can be used up relatively quickly.

Most large supermarkets offer a wide variety of vinegars but it is also worth remembering that wine merchants sometimes sell vinegars from small producers. Delicatessens, organic shops and markets, wholefood shops and speciality food stores are all great places to discover vinegars from around the world.

Organic vinegars

Vinegars produced to organic standards are widely available, especially cider vinegars. Some unpasteurized organic vinegars may contain live bacteria that could develop a mother of vinegar once the contents of the bottle are part used, exposing the surface of the

Above: Different vinegars have been produced all over the world for centuries. The ingredients and method of production influence the overall quality of each vinegar.

Above: Taste one or two types of vinegar by sipping a small amount from a spoon.

Above: Dip small cubes of white bread into vinegar to compare several types.

Above: Vinegar can be made from different fruit or vegetables.

vinegar to air. This could be useful for anyone wanting to try making their own vinegar.

Tasting tips

Take advantage of vinegar tasting sessions offered by wine merchants, supermarkets or other specialist stores. It is a very good way of sampling and comparing different brands and qualities of the same type of vinegar.

It is also worthwhile comparing a couple of different vinegars at home. Sipping a little off a teaspoon is the obvious way of tasting, drinking plain water between samples, but it is better to dip small cubes of good-quality, close-textured white bread into a small amount of vinegar. Make the cubes similar in size and without crust. Pour a little vinegar on a saucer and lightly dip the bread in it. This is a good way of comparing the sharpness and additional flavours in different vinegars. Drink water between samples and compare only two or three types at once to avoid confusion.

Make a note of impressions on a label, with ideas for using the vinegar, then tie it around the bottle. Experiment with the suggestion soon afterwards to confirm or correct the first impression.

Alternative ingredients

While the majority of commercial vinegars are based on wine, ale or a fermented fruit or grain base, all sorts of different ingredients are used for small-scale production. These vinegars are not generally available in the average supermarket but independent delicatessens, wholefood or healthfood stores and mail order specialist suppliers are the best places to find and sample the more unusual end of the vinegar spectrum.

Millet, wheat, sorghum and oats may all be used in the same way as rice or barley. The sap of palm trees is also used in vinegars, particularly in the Philippines. A wide variety of vegetables, including potatoes, beetroot (beet), carrot, asparagus and cucumbers are also used for making specialist brewed vinegars.

STORING VINEGAR

All varieties of vinegar should be stored in an airtight bottle in a cool, dark place. Those vinegars with live bacteria remaining will begin to ferment if they are not stored in this way.

Pasteurized or distilled vinegars will keep indefinitely, although the flavour will lose strength. However, it is generally a good idea to use all vinegar by the date suggested on the label.

If vinegar smells bad or rotten, it should be discarded.

BALSAMIC VINEGAR

Probably one of the most famous of vinegars, 'aceto balsamico' originates from Northern Italy, where it was first fashionable among wealthy 18th-century households.

Balsamic vinegar has been produced in the Italian towns of Modena and Reggio Emilia since the Middle Ages. Barrels of the precious 'balsamico' were originally stored in lofts, where they fermented over some years, then matured for more years, warming up in summer and cooling in winter, gradually evaporating and becoming more concentrated in flavour.

The vinegar is fermented from a must of Trebbiano grapes. Harvested from the Modena or Reggio Emilio areas of Northern Italy, the cooked must is reduced to a dark syrup before it is ready for fermentation using a mother reserved from previous batches. The process of fermentation from the fruit, through alcohol to acetic acid, takes some three years. In the first stage the sugar is converted to alcohol before the vinegar-producing bacteria take over. When the fermentation is complete, then maturation begins. This is no quick process but something that takes many years – from 12 years and upwards, to 50 years or many more for the greatest (and most expensive) vinegars made by this traditional method.

During the process, the vinegar is carefully transferred through a whole series of barrels. The barrels are made from different woods, to impart their character at various stages. It is said that spices may be added at different stages but such ingredients are closely guarded secrets and the barrels and mother of vinegar become treasures that gradually work their magic on successive batches of vinegar. As some vinegar is drawn off the final smaller

Above: Real 'tradizionale' balsamic vinegar is expensive and should be bought in small quantities and used sparingly.

barrels, so younger stock is moved along and some new vinegar is added to the early barrels to begin its ageing process. This amazing chain of ancient barrels, gradually reducing in size, is preserved and kept working by those who practise traditional balsamic-making methods.

Genuine balsamic vinegar made by this long process is expensive and sold in small quantities. It will be labelled as 'Aceto Balsamico Tradizionale di Modena' or 'Aceto Balsamico Tradizionale di Reggio Emilia', depending on exactly where it was produced.

Balsamic vinegar has been made in this way in Modena and Reggio Emilia since the Middle Ages, and vinegars bearing these labels are protected by the Italian quality assurance label

Above: Traditionally produced balsamic vinegars from Modena and Reggio Emilia are aged for at least 12 years.

Above: Methods and ingredients are passed down through generations of vinegar producers and are a strict secret.

Above: There are many varieties of balsamic vinegar. Check labels for details of ingredients and production methods.

Above: In Italy, rich balsamic vinegar is traditionally served as a dip for fresh white bread before a meal.

Above: Balsamic vinegar is often used in cooking to deglaze pans after frying or to bring sharpness to a sauce.

'Denominazione di origine controllata' and the European Union's Protected Designation of Origin (PDO). Other balsamic vinegars produced which are not long-aged and do not undergo such strict quality control are simply known as 'aceto balsamico'.

There are all sorts of balsamic vinegars available and the quality varies widely. However, even though less expensive varieties may not be 'proper' balsamic vinegar and nowhere near as fantastic as the traditional, long-aged vinegar, it does not mean they should be dismissed. Think of them as 'balsamic-style' vinegars or vinegar dressings and instead of comparing them in a negative sense with their original source of inspiration, taste them as products in their own right. Some of them are good while others are deeply disappointing. Price is a first guide: a big bottle for very little money is unlikely to be high quality. Check the precise description of the vinegar and the ingredients list to decide whether the bottle contains balsamic vinegar or a sweetened and flavoured vinegar. For example, the product may be a wine vinegar with grape concentrate or a wine vinegar with sugar, fruit juice and caramel.

USE

- As a dip or condiment for bread.
- As a rich dressing, applied by sparing drizzles to individual foods, either savoury or sweet, including fruit.
- For deglazing cooking pans, for example after frying meat or roasting vegetables.
- To enrich sauces and gravies.
- Mixed with water and demerara (raw) sugar as a glaze for meat such as pork or beef.
- To brush over chicken or fish before grilling (broiling)
- In salad dressings.
- To flavour mousses, creams and pâtés.
- To sharpen or enrich drinks, for example with blackcurrant or cranberry juice, topped up with mineral water.
- In some cocktails.
- In smoothies, with fresh fruit, such as strawberries or raspberries.

WHITE BALSAMIC VINEGAR

White balsamic vinegars are a good example of products that are not balsamic vinegar at all: check the label and you may find the contents referred to as 'balsamic vinegar condiment'. There are lots of versions on sale of different qualities and prices.

White balsamic vinegar is a mixture of white wine vinegar and white grape must, and does not undergo the same lengthy ageing process as traditional dark balsamic vinegar.

WINE VINEGAR

Red or white wine vinegar, produced by fermenting wine, varies in flavour and quality according to the type of wine, the fermentation process and whether the vinegar has been matured.

Do not automatically assume that all wine vinegars are superior just because they are made from wine – cider vinegar is often a better 'everyday' alternative to inexpensive wine vinegar that can be too harsh and lacking subtle flavours.

Standard, mass-produced inexpensive wine vinegars are unlikely to vary in subtle flavours from one to another, but it is definitely worth sampling different brands, especially among the slightly more expensive ranges, to find one that you prefer.

White and red wine vinegars

All white wine vinegars are light and crisp, and some are very sharp. Red wine vinegars are pink to light red in colour and some are matured, with a stronger, fuller flavour.

Above: Wine vinegar is the most commonly used vinegar in the Mediterranean.

Above: Some wine vinegar is made from a specific wine, such as pinot noir vinegar, or Chardonnay vinegar.

Vinegars made by traditional slow methods and matured have more complex flavours, and their piquancy is rounded or balanced slightly by the fruity tones of the original wine. Sometimes, the depth of flavour comes from the barrel in which the vinegar was aged.

Look for wine vinegars that have been made by the Orléans method, which involves the traditional processes of slow fermentation and maturing. Named wine vinegars will hint at the characteristics of the wine from which they have been fermented. Among the reds, Merlot, Cabernet Sauvignon and pinot noir vinegars are all available. Look out for the wine vinegars of different countries to broaden the tasting experience.

Above: White wine vinegar may be used to sharpen cooking liquor and balance sweet flavours.

USE
- In hot and cold sauces and in salad dressings.
- As a marinade ingredient for white fish or sousing liquid for oily fish.
- In soups, casseroles or stews.
- To sharpen sauce mixtures or drinks.
- For making preserves (it is usually a waste to throw high-quality wine vinegar into a general vegetable pickle or other mixed preserve but less expensive products are suitable).
- For making flavoured vinegars, for example by adding herbs, fruit or spices.
- For home-made natural skin and hair treatments.
- For cleaning and restoring wood (less expensive wine vinegars are most useful for this purpose).

Champagne vinegar

This is made from Champagne and also known as Reims vinegar (vinaigre de Reims), named after the town of Reims, which is centrally located in the Champagne region of France. All Champagne vinegar should be produced here – so check the label.

Champagne vinegar is aged in oak barrels for one year. As a wine vinegar, it is quite sharp but lighter in flavour than white or red wine vinegars. Although, in theory, it should be a superior vinegar, it is worth remembering that its quality depends on the original quality of the wine. A good Champagne vinegar will hint at the pinot noir flavours that are so characteristic of the original wine complemented by the maturing process used for the vinegar.

It is not so long ago that Champagne vinegar was a treat available from specialist shops and wine merchants but it now features among the own-brand products of larger supermarkets.

Above: Champagne vinegar is aged in huge oak barrels, and must be produced in the Champagne region of France.

USE

- For marinades, dressings and sauces, both savoury and sweet.
- In light dressings or to deglaze the cooking pan for fish and shellfish, especially salmon.
- As an accompaniment to oysters.
- To sharpen drinks and desserts, and bring a hint of piquancy to fresh ripe and sweet berries.
- For making delicately flavoured herb, fruit or flower vinegars.
- Drizzled over very sweet ripe peaches.
- To sharpen rich fruit salads or compotes, especially those with dried fruits.
- With nut or truffle oil to make a sublime vinaigrette.

Sherry vinegar

Spain is known for the quality of its sherry vinegars (vinagre de Jerez). Like other wine vinegars, the quality, or blend, of sherries dictates the quality of the vinegar. Good sherry vinegar that is well matured has its sharpness balanced by the full and distinct flavours of the original wine. The best can be mellow

Above: The best quality Champagne vinegars will retain some of the taste of the original Champagne wine.

Above: The colour of sherry vinegar varies from golden through to rich dark brown.

and rich, and compared by some to balsamic vinegar. However, as for other vinegars, the flavour depends on the individual product and there are many sherry vinegars available.

USE

- In marinades for red meat, rich poultry and game.
- For bringing a hint of piquancy to rich gravies and sauces, or deglazing cooking pans after frying meat or oily fish.
- For rich salad dressings or cold dressings for meat.
- In soups and pasta sauces.
- In preserves such as pickled shallots.
- Drizzled over grilled (broiled) halloumi cheese or for dressing and marinating cubes of feta or manchego cheese.
- To bring a hint of contrasting piquancy to rich fruit compotes, chocolate syrups, caramel sauces, praline or toffee mixtures.
- For making rich and warming, throat-clearing hot toddies, especially with honey.
- For flavouring rich ice creams, such as brown bread or nut ice creams.

MALT VINEGAR

Produced in Britain and in other European countries with a history of ale brewing, malt vinegar that is available today is derived from the traditional product 'alegar', or ale vinegar.

Barley, the basic ingredient of malt vinegar, is sprouted, a process that leads to starch being converted to maltose, a type of sugar (not as sweet as sucrose) that develops when starch is digested. Rich, full-flavoured malted barley is fermented to produce alcohol and then onwards to make vinegar. Caramel is added to intensify the flavour and colour of the vinegar. It is important to check the label to ensure that the product is malt vinegar and not an inferior, bitter mixture of diluted acetic acid and caramel or other colouring and flavouring ingredients.

Malt vinegar is famous as the condiment sprinkled on British fish and chips. While some may be horrified at the idea, a little good malt vinegar perfectly complements crisp, hot chips (French fries) and white fish in crunchy batter. Wine vinegar is either too crisp and tart

or fruity to cut the flavours to perfection. Malt vinegar is also useful for pickling and making rich chutneys and ketchups. Malt vinegar is sold ready spiced for pickling vegetables, such as onions.

USE

- As a table condiment.
- For making chutneys, pickling and in other savoury preserves.
- As a base for sauces and marinades.
- For household and outdoor use.
- As a general souring agent in cooking, for example in sweet-sour sauces and dishes.

Distilled malt vinegar, white vinegar or spirit vinegar

Distilled malt vinegar is a concentrated malt vinegar with a higher acid content, which makes it ideal for

preserving vegetables and other ingredients with a high water content. Water is likely to dilute pickles, making them more vulnerable to attack from micro-organisms. When caramel is not added the distilled vinegar is clear and it may be called white vinegar or distilled white vinegar. The flavour is similar but not as full.

It is important to avoid cheap alternatives that are made by diluting acetic acid with water as they do not have the same flavour as real distilled malt vinegar. Check labels carefully for ingredients – these should consist only of the vinegar and not of acetic acid plus other items. 'Non-brewed condiment' is a term that can be used for this diluted acetic acid.

Above: Salt and malt vinegar – the classic accompaniment to British fish and chips.

Above: Distilled malt vinegar can be dark brown, or clear if no caramel is added.

Above: Distilled white vinegar is most economical to use around the home.

USE

- As a pickling vinegar, especially for eggs or light vegetables, such as red cabbage.
- As a raising agent in baking.
- For first aid – such as treating bites or stings.
- For household and outdoor use.

Light malt vinegar

This is a pale gold-coloured vinegar brewed as for ordinary malt vinegar but without the addition of caramel. The flavour is milder and very similar in taste to rice vinegar.

USE

- As a pickling vinegar when a milder flavour is required.
- For salad dressings and sauces.
- In marinades for meat, poultry or fish.
- To sharpen drinks.
- As a raising agent in baking.

Pickling vinegar

This is distilled malt vinegar flavoured with a mix of spices typically used for pickling vegetables. It may be dark malt vinegar or white vinegar, but in

Above: Light malt vinegar has a milder taste than ordinary malt vinegar.

Above: Onions are often pickled in malt vinegar which has been spiced with peppercorns, cloves and chillies.

both cases the flavour of the spices is usually very mild. Black peppercorns, coriander seeds, ginger, allspice, cloves, mustard seeds, dried red chillies and cinnamon may all be used to spice vinegar for pickling. To impart a more distinct flavour when pickling, it is best to infuse the whole spices in plain malt vinegar. When spicing your own vinegar at home, malt, wine and cider vinegars can all be used.

USE

- In light fruit chutneys (such as peach, pear or apricot) or for making fruit or herb jellies, such as apple or mint jelly.
- For pickling onions, red cabbage, chillis, cucumbers, beetroot (beet) or hard-boiled eggs.
- To make sweet fruit pickles (sugar should be dissolved in the vinegar to sweeten it slightly).
- In sweet-sour braised vegetables, such as red cabbage, when vinegar is lightly spiced.
- In punchy salad dressings, for example to dress meat salads.

Above: Ready-spiced pickling vinegar is available to buy, but it is very easy to make using malt, wine or cider vinegar.

MAKING SPICED VINEGAR

Place 1 cinnamon stick, 12 cloves, 60ml/4 tbsp coriander seeds, 30 ml/ 2 tbsp mustard seeds and 15 ml/ 1 tbsp black peppercorns in a large heavy pan.

Place over a medium heat, shaking the pan often, until the spices are lightly roasted and giving off their aroma. Do not overcook the spices until they begin to pop or brown. Add 4 dried red chillies and pour in a little vinegar taken from 2.4 litres/4 pints/10 cups malt vinegar, adding enough to cover the spices generously.

Bring to the boil, then remove from the heat and pour in the remaining vinegar. Cover tightly and leave to infuse for 1–5 days. Strain the vinegar into clean bottles and cover tightly, then store in a cool place.

CIDER VINEGAR

One of the oldest and original cottage-industry vinegars, apple cider vinegar is still renowned for its health benefits; it is also one of the most useful of culinary vinegars.

One of the traditional vinegars made in apple-growing areas, and well known especially in the US and Great Britain, cider vinegar is an ancient cure-all that has retained its reputation as a healing agent and general health-giving potion. Cider vinegar is said to make a positive contribution to everything from weight-loss and relieving arthritis to reducing cholesterol levels and the prevention of heart disease. It is available to purchase in pharmacies and healthfood shops, as well as in supermarkets. Many people swear by a daily dose of cider vinegar for general wellbeing, while others employ it as a beauty product. It is also one of the most versatile of culinary vinegars.

Above: Cider vinegar varies in colour, from colourless or very pale yellow to a rich gold, and can be cloudy or clear.

Above: The best varieties of cider vinegar are made from just one type of apple.

Above: Cider vinegar is prized for its health benefits. Drinking a teaspoon a day is said to help all manner of ailments.

Cider vinegar is made from apples, starting with the raw fruit, fruit juice, apple wine or cider. There are delicate, yet distinct apple-like flavours in some cider vinegars. The best, and most expensive, vinegars are made from single types of apples. Unsurprisingly, the quality of the base product influences the quality of vinegar and there is a very large selection from which to choose. Cloudy vinegars that contain some fruit residue may not look as appealing, but they are often superior in taste. There are both organic and unpasteurized cider vinegars, made by fast or slow methods, from small or large producers. The choice can be overwhelming. Read the labels on

cider vinegar bottles to check if there are any added ingredients or if the product is an inferior blend of vinegar or acetic acid with apple juice.

As a general rule, cider vinegar is milder than other wine and malt vinegars; some are clear, others cloudy, and the colour of the different types varies from pale yellow to a honey-like gold.

CIDER VINEGAR STARTER

Many varieties of live organic cider vinegar are available. These can be used as a starting point for experiments with making vinegar at home. It should be added to live organic cider or wine.

Live organic cider vinegar

When using cider vinegar for its health benefits, it is important to select an organic product, which is made from the fruit (rather than vinegar plus juice), preferably live, without any added chemicals to 'kill off' the vinegar bacteria.

Cider vinegar must be stored in an airtight container in a cool, dark place to preserve any vitamin C it may contain. Buy modest quantities rather than large amounts so that it will be used within the date indicated on the bottle.

As well as for internal health benefits, the fruit acids and residues in this vinegar are helpful in home-made beauty products, for example as an exfoliator to remove dead skin residue and as a face freshener.

Diet and cider vinegar

Cider vinegar has been associated with weight-loss, with claims that drinking it helps to reduce fat absorption or even break down fat in the body. Cider vinegar tablets or capsules are sold in healthfood stores as a natural slimming

Above: On its own, or with other ingredients, cider vinegar can be used to make natural home remedies.

aid. While there is no scientific evidence to support the digestive theories, including cider vinegar as part of a balanced diet is a good idea, especially if it encourages a taste for less-sweet, less-rich foods and when it is used in beverages that replace high-sugar drinks.

Above: Using cider vinegar on the skin after bathing can help to redress any imbalance in the skin's pH levels.

USE

- As a home remedy or preventative agent for promoting good health.
- For first-aid, such as treating insect bites or stings.
- To soothe tired muscles and aches and pains by adding a little to a warm bath.
- For home-made beauty treatments such as hair rinses and face packs.
- As a base for home-made vinegar.
- As a cleaning agent around the home.
- In salad dressings, sauces and dips.
- In relishes and salsas.
- To bring piquancy to a wide variety of savoury or sweet dishes without making them too harsh.
- In sweet dishes such as sorbets, ice creams and mousses.
- For drinks and cordials.
- In smoothies – try a quick cider vinegar and banana smoothie by whizzing up a banana and yogurt in a blender then drizzling in a little cider vinegar for a zesty flavour.
- In some preserves and pickles.
- As a base for home-made flavoured vinegars, particularly herb vinegars.

Above: The mild taste of cider vinegar makes it a very good base for home-made herb-flavoured vinegars.

Above: Cider vinegar is often used to pickle ingredients with a strong flavour, such as garlic or shallots.

RICE VINEGAR

An essential part of Asian cuisine for centuries, many types of rice vinegar and rice vinegar dressings are now available in specialist and healthfood stores as well as supermarkets.

Rice vinegars may be fermented either from rice or from rice wine. Traditional methods started with the process of fermenting the rice itself, rather than making rice wine into vinegar. Rice vinegars vary in colour and flavour; some are milder than wine vinegar or malt vinegar while others are harsher and more comparable with light malt vinegar. Some of the rice vinegars available are sweetened or flavoured and coloured, so it is worth checking the ingredients on the label carefully.

As well as the familiar Asian rice vinegars, Californian rice vinegars are also available. They are made from grain harvested locally and include organic brown rice vinegars, which can be plain or seasoned. Brown rice vinegar can have a very

strong and distinctive flavour, which is more noticeable than that of the majority of wine or malt vinegars.

Chinese rice vinegars range in colour from clear to various shades of red and brown, through to black. In Chinese medicine, rice vinegar is used in soup that is recommended to ease asthma. It is also thought to be clearing, helpful for good digestion and useful for promoting liver and stomach function. It is also believed to help external conditions of the skin.

As a rule, Japanese rice vinegar, or su, is lighter than Chinese vinegar. It is available in many varieties and qualities. Traditionally brewed vinegars made from unpolished glutinous rice have a reputation for good quality. Notice the difference between rice vinegar and rice vinegar dressing, as

Above: Rice vinegars vary in colour, from clear, through to amber-yellow and red.

the latter will be seasoned and sweetened or flavoured, for example, ready for mixing with sushi rice. Vinegared rice, or su-meshi, is made by dressing cooked rice with a mixture of rice vinegar, sugar and salt.

Rice vinegar may be flavoured with Japanese citrus fruit, such as dai dai, a type of bitter orange. Ready-made, this sauce is sold as ponzu in Japan.

Black rice vinegar

Popular in southern China, black rice vinegar has a full flavour and may be enriched with malt. Black vinegar may be aged and intense in flavour. Chinkiang vinegar, which originated in the city of Zhenjiang in the eastern coastal province of

Above: Black rice vinegar is popular in Japan, where it is sold in health drinks.

Above: Bottles of Chinkiang vinegar are checked on the production line.

Above: A wide variety of rice vinegars can be found in Asian supermarkets. They vary in taste and quality.

Jiangsu in China, has a reputation as being one of the best. The longer the vinegar is kept, the stronger its aroma becomes. As with other superior varieties of vinegars, there are always inexpensive alternatives of inferior quality, so it is worth experimenting to find a good product.

Above: Japanese rice vinegar is an essential part of sushi preparation.

Above: Black, red and white rice vinegars are often used in dipping sauces in Asian cuisine.

In Japan, black rice vinegar (kurozu) is prized for its health benefits, and in recent years has been sold in vinegar drinks which are available from supermarkets and vending machines, as well as from dedicated vinegar bars.

USE

- In meat dishes as a marinade and tenderizing agent.
- As a dipping sauce.
- As a substitute for balsamic vinegar.
- In soups and sauces.
- In stir-fries.
- In health drinks.

Red rice vinegar

This varies from a lighter variety of black rice vinegar to a seasoned, sweet and full-flavoured vinegar that is popular as a dipping sauce or for adding to soups.

USE

- As a substitute for black vinegar – just add a little sugar.
- As a dipping sauce.
- In noodles, soup and seafood dishes.

White rice vinegar

This is colourless or a very pale yellow colour, light in flavour, with different regional variations or specialities.

USE

- To dress vegetables and salads.
- In sweet-and-sour dishes or hot-and-sour recipes, such as the classic soup.
- For making refreshing drinks.
- For dressing rice and in Japanese sushi.
- As a pickling medium, for example, for garlic or ginger.
- In Western cooking, in salad dressings, sauces and drinks.
- In both hot and cold sauces, including dipping sauces and sweet-and-sour sauces.

MAKING SUSHI RICE

Vinegared rice (su-meshi) is the essential base for all kinds of sushi. Put 200g/7oz/ 1 cup Japanese short grain rice in a large bowl and wash in plenty of water, until it runs clear. Tip into a sieve (strainer) and leave to drain for 1 hour.

Put the rice into a small, deep pan with 15 per cent more water, i.e. 250ml/8fl oz/1⅕ cups water to 200g/7oz/1 cup rice. Cover and bring to the boil. This takes about 5 minutes. Reduce the heat and simmer for 12 minutes without lifting the lid. You should hear a faint crackling noise. The rice should now have absorbed the water. Remove from the heat and leave for 10 minutes.

Transfer the cooked rice to a wet Japanese rice tub or bowl. Mix 40ml/ 8 tsp rice vinegar with 20ml/4 tsp caster (superfine) sugar and 5ml/1 tsp salt, until dissolved. Add to the rice, fluffing with a wet spatula. Cover the bowl with a wet dish towel and leave to cool.

OTHER VINEGARS

Vinegar can be made from many ingredients. Fermented fruit vinegars are popular around the world, while coconut, cane and palm vinegars are used predominantly in South-east Asia.

Vinegar is a global ingredient, used in the cooking of almost every cuisine, as well as for health, beauty and household purposes. It has traditionally been made from whatever ingredients are grown locally.

Fermented fruit vinegar

As well as wine and cider vinegars originating from grapes and apples, many other varieties of fruit are involved in the vinegar fermentation process. Vinegars made from fermented fruit are different from fruit-flavoured vinegars, where fruit are macerated in a vinegar to impart their flavour to it. Fermented fruit vinegars are expensive, and usually produced in relatively modest quantities. A wide variety of fruit may be used, including familiar raspberries, but also tomatoes, blueberries, pears, quinces, pineapple, bananas, peaches or apricots. Plums, fresh or dried, are also used. Many of these vinegars will retain the flavour of the original fruit. Most fresh fruit vinegars are produced in Europe, but persimmon vinegar is often used in Korean cooking (as well as apple-based cider vinegars).

As well as fresh fruit, dried fruit, such as raisins, currants, dates or figs, can be used to make richly flavoured vinegars. Raisin vinegar is used in Middle Eastern cooking and is a cloudy brown colour, with a mild taste. Dried fruit vinegars are predominantly produced in countries such as Turkey, Greece, Spain, Morocco and Algeria.

Above: Vinegar fermented from raisins is popular in the Middle Eastern countries, where it is used in cooking.

The best fermented fruit vinegars are aromatic, carrying the distinct flavour of the fruit in a well-balanced vinegar that is acidic and fruity but not sweetened. These products are expensive and a vinegar treat rather than a store-cupboard staple.

USE
● As dressings for individual ingredients – light fruit vinegars are best for this.
● In marinades for meat, poultry or fish.
● To dress fresh fruit.
● In desserts and drinks.
● Drizzled over light cheeses, such as mozzarella or soft goat's cheese.
● To enrich savoury sauces or casseroles. Rich fruit vinegars may be paired with meat dishes.
● With nut oils for rich salad dressings.

Above: Fig vinegar, as well as vinegar fermented from raisins and dates, is used in Mediterranean countries.

Above: Raspberry vinegar has a distinctive red colour and can be quite sweet if sugar has been added.

Coconut vinegar

This is made from fermented coconut water (the clear liquid naturally occurring in coconuts, not to be confused with coconut milk, which is made from the flesh of the fruit). Coconut vinegar is very popular in South-east Asian cuisine, particularly in the Philippines. It is a cloudy white vinegar, with a slight flavour of yeast.

In Filipino cooking, ingredients are often marinated in coconut vinegar, ginger and garlic before cooking, and raw ingredients such as oily fish and shrimp are cured in lime juice and coconut vinegar, as is exemplified in the national favourite kinilaw (Filipino cured herring).

The Filipino penchant for sweet and sour notes is achieved by combining coconut vinegar or kalamansi lime juice with palm sugar (jaggery) or cane sugar, as in adobo (chicken and pork cooked with vinegar and ginger), the national dish, which originally hailed from Mexico.

Above: In the Philippines, coconut vinegar is used extensively. It can be found in speciality stores.

USE
- In South-east Asian dishes, such as stews, for authentic flavour.
- As a marinade for meat, poultry and fish dishes.
- In sauces and soups.
- To cure fish and shellfish.

Above: Filipino cured herring is marinated in coconut vinegar mixed with lime juice, and flavoured with ginger and chillies.

Cane vinegar

Made from fermented sugar cane, this is popular in the Philippines. The colour varies from pale yellow to dark brown, and it has a mild flavour, not unlike rice vinegar. There is no trace of sweetness in the vinegar from the sugar cane.

USE
- In salads and sauces, or as a glaze for cooked meats.
- In vinaigrettes when combined with rich nut oils such as walnut oil.

Palm vinegar

This cloudy white vinegar is made from the fermented sap from the fruit of the nipa palm. It looks similar to coconut vinegar, but has a milder taste. Also popular in the Philippines, it is milder than wine or cider vinegars.

USE
- In dipping sauces with chilli and garlic.
- As a dressing for salads and vegetables.
- As a marinade for meat, poultry or fish.
- In soups and sauces.

Above: A sugar cane farmer in the Philippines tests the sugar cane juice which is fermenting in earthenware jars.

Above: The fruit of the nipa palm produces a sap which is fermented to make alcohol and palm vinegar.

VINEGAR FOR HEALTH AND HEALING

This section provides a selection of ideas that make the most of vinegar's health-giving properties. Cider vinegar is particularly appreciated in a wide variety of uses for promoting positive health and healing: from an antiseptic agent or external coolant to an internal provider of holistic help for long-term wellbeing. A little vinegar may be taken regularly as a refreshing ingredient in health-promoting drinks. It is also a great standby for first aid use when a simple home cure is sufficient.

Left: Alongside a healthy diet and regular exercise, vinegar contributes to general good health and wellbeing and is used to treat specific ailments.

VINEGAR AS A TONIC

Before modern antiseptics, drugs, vitamin pills and medicines became everyday potions, vinegar, especially cider vinegar, was widely used to promote good health.

Vinegar has been used for its health benefits for centuries. The positive effect that vinegar is said to have on a range of chronic diseases continues to be supported. Many people assert that their symptoms are reduced by taking regular doses of vinegar. Cider vinegar in particular has been taken for many years as a general health tonic as well as to treat specific ailments. However, modern medical theories cannot offer a precise explanation for why and how vinegar works. The general nutritional benefits and its influence on eating patterns are likely factors, but there is greater evidence for the dietary role that vinegar can play in supporting the acid-alkali balance in the body.

WHICH VINEGAR TO USE?

Organic cider vinegar is generally recommended for remedies and as a tonic. There are also different flavoured organic vinegars based on cider vinegar, including raspberry vinegar, that are excellent for drinks and dressings. It is also worth looking out for natural, live vinegars fermented from fruit, such as raspberry and blueberry vinegars, as they provide the goodness from the particular type of fruit, just as cider vinegar does from apples.

Cider vinegar is also the recommended choice for external treatments, for example it can be used for treating mild skin conditions as well as for making pleasant rinses and washes, but other types, such as malt or rice vinegar, are also useful for their health benefits.

Above: Vinegar can make a positive contribution to health when part of a balanced diet.

Ancient harmony and health

Vinegar was prescribed in ancient times to support natural harmony or balance. The ancient Greek physician Hippocrates recommended vinegar for clearing and cleansing the system and functions associated with the kidneys and liver. Vinegar is one of many ingredients still used in Chinese medicine for promoting good health through diet. It is also believed to help arrest bleeding, aid digestion, combat parasites, treat jaundice, skin complaints and nose bleeds.

Vinegar and nutrients

The nutritional value of vinegar depends on the ingredients from which it is made. Typically, vinegar contains minerals, especially potassium, as well as some chloride, phosphorus, magnesium, sodium and sulphur, with a little calcium and traces of other minerals. Minerals are important in modest amounts and are widely distributed among foods. Although the proportions are not high in vinegar compared to the main ingredients used in meals, vinegar

Above: In 1950s America, a mixture of vinegar and honey called 'Honegar' was marketed as a cure-all elixir. There are still many people who swear by the healing powers of vinegar.

makes a helpful contribution when it is used regularly. Vinegar is a valuable flavouring and seasoning ingredient, in cooked dishes, dressings, sauces and drinks.

The acid-alkali balance

For healthy, normal function the body has to maintain an important acid-alkali balance. Gastric juices are acidic so that they can digest food, but the acids must not enter the digestive system. The body has an efficient buffer system that keeps the cells at the required levels. Deficiencies and health problems are often reflected by changes in the pH levels in the body and these have to be rectified, both in the short term and by long-term control. Eating the right diet helps to rectify poor balance because some foods are acid-forming while others are neutral or alkaline-forming. Vinegar is a neutral to slightly alkaline-forming food.

Whether a food is acid- or alkaline-forming depends on its nutritional make-up and what effect it has during digestion. For example, acid-forming foods include fish, meat and eggs, dairy products, refined starches and sugary foods. Fruit and vegetables are generally alkaline-forming. Eating 80 per cent alkaline-forming and 20 per cent acid-forming foods is recommended as a general rule. Eating foods that create an acidic environment encourages the body to secrete alkaline substances into the blood, maintaining the balance.

The power of cider vinegar

This has long been regarded as a 'cure-all'. Natural, unpasteurized, cider vinegar made by traditional methods is thought to be beneficial for the pectin it contains from the apples. Pectin, a form of soluble fibre also found in other ingredients, such as oats, can help to maintain healthy blood cholesterol levels.

Consumed regularly, there is much evidence that vinegar (in particular cider vinegar) helps to relieve joint pain associated with arthritis. Vinegar also helps to promote a healthy cardiovascular system, including good blood cholesterol levels and blood pressure.

Vinegar is useful for protecting against general colds and helping to relieve cold symptoms. It promotes good digestion, helps to reduce and avoid indigestion and to relieve diarrhoea. As a short-term tonic it can help to cleanse the system as part of a process of detoxing intended to promote good liver function.

It is also useful as a general tonic for promoting a sense of wellbeing. Warm cider vinegar drinks can help to promote sleep.

Anyone suffering from a medical condition should take professional medical advice before introducing new foods into the diet or making any diet changes which are intended to alleviate symptoms. Aside from using vinegar as a home remedy for minor conditions, for anyone who is active and healthy, using cider vinegar as a tonic or everyday ingredient in drinks or daily cooking is a great way of promoting good health. To be effective, regular use of cider vinegar should be part of a general diet and lifestyle review that should be planned and seen as a long term change.

Above: Vinegar can be taken internally to help a variety of complaints.

VINEGAR FOR HEALING

A small bottle of vinegar makes a handy addition to any first aid kit or bathroom cabinet. It can be used on its own or with other ingredients to treat many common ailments.

The natural acidic and astringent properties of vinegar mean that it can be effective as a disinfectant, antiseptic and coolant. It is a good idea to keep a small bottle of vinegar in the bathroom, where it will be useful for treating a variety of complaints. It is very good for cleansing cuts and scrapes in the first instance.

External use of vinegar

Cider vinegar contains alpha-hydroxy acids, or natural fruit acids, which are very good for the skin. It makes a fantastic natural exfoliator, on its own or mixed with other ingredients. It can be used in washes and rinses or applied directly to skin, either neat or diluted, to calm and cleanse skin conditions and infections, including acne and blocked pores.

Cider vinegar is a good choice for treating skin problems, but other types of vinegar, in particular malt and rice vinegar, are also effective. They are useful neat or diluted as astringents, as well as an antiseptics for clearing simple outbreaks of spots. A little neat vinegar on a cotton wool ball can be applied to spots to clean the area and prevent further infection.

Vinegar is a simple and effective antiseptic for cleaning up minor cuts and grazes, and can be used for cooling and cleansing insect bites, wasp stings and jellyfish stings. In parts of northern Australia where box jellyfish are common, vinegar is even provided on beaches so it can be applied quickly to stings while

Above: Bottles of vinegar form part of an Australian lifeguard's equipment – vinegar is used to treat jellyfish stings.

waiting for medical treatment. The acetic acid in vinegar is said to disable any venom that has not yet entered the bloodstream. Box jellyfish stings can be fatal to humans, so proper medical attention must be sought.

As well as treating their bites and stings, vinegar can also be used as a repellent to keep insects, including mosquitoes, away. Transferring a small amount of vinegar to a spray bottle is a great way of creating a natural insect spray that will not be harmful to children or pets.

Vinegar is helpful for clearing infestations of head lice and for resolving mild dandruff problems. For healthy feet, vinegar can be used

Above: Cider vinegar can be applied directly to the skin to treat skin conditions.

Above: Vinegar is a natural antiseptic. Use to soothe and calm minor irritations.

Above: The natural properties of vinegar mean that it has been used for many centuries as a disinfectant, antiseptic and general health tonic.

in the removal of corns or for overcoming common fungal infections, such as athlete's foot. It is also good for reducing soreness and swelling and for soothing dry or itchy skin.

Vinegar baths or soaks can help to relieve tired and aching muscles, and to reduce the soreness of bruising, as well as to cool sunburn. Minor burns can be treated by dabbing on a little neat vinegar.

Vinegar steam baths are useful for clearing a blocked-up nose and other symptoms of a head cold. Mixed with honey, another natural antiseptic, and different herbs, vinegar makes a very effective gargle to soothe a sore throat.

Vinegar in natural remedies

Use simple homely treatments for minor ailments that do not pose any significant threat to health.

For example, many of the vinegar remedies suggested here are perfect to soothe the symptoms of common illnesses such as coughs and colds.

Wellbeing is defined as part of good health. Some of the vinegar remedies and drinks in this book are very useful for anyone who is run down or stressed. Such simple remedies are often sufficient to prevent a condition from developing. It is important to recognize progress and relief of symptoms or acknowledge the need for medical help.

Make any changes gradually and they are more likely to be enduring and effective. With a diagnosed condition or illness, you should never dismiss medical advice and entirely reject prescribed drugs. It is important to differentiate between vital treatment and complementary, holistic and dietary changes with the help of a doctor.

VINEGAR DRINKS FOR HEALTH

These vinegar drinks are refreshing, reinvigorating and restoring. Simply add a spoonful of vinegar to a glass of water, or mix with other ingredients.

Indigestion

Cider vinegar contains pectin, which is good for digestion. It also acts as an antiseptic in the intestines.

Add 5ml/1 tsp cider vinegar to a small glass of water (about 175ml/6fl oz/ ¾ cup) and mix well. Drink this daily to relieve indigestion.

Asthma

In Chinese medicine, vinegar is one of the ingredients believed to help treat asthma. Vinegar is used in drinks and in soup (for example hot-and-sour soup).

Add 5ml/1 tsp vinegar to 250ml/8fl oz/ 1 cup of hot water. Drink daily to help with the symptoms of asthma.

Stress

Fennel, vinegar and lemon tea makes a good mid-morning stress buster or afternoon pick-me-up. Fennel tea is available from health stores.

Add 5ml/1 tsp cider vinegar and a squeeze of fresh lemon juice to a cup of hot fennel tea.

Tension

Camomile tea, apple juice, cider vinegar and honey makes a soothing drink.

Add 5ml/1 tsp each of cider vinegar and honey and 15ml/1 tbsp apple juice to a cup of calming camomile tea.

Insomnia

A cider vinegar and honey drink before bed helps to promote calm and can help combat insomnia.

Mix 10ml/2 tsp cider vinegar with 10ml/ 2 tsp honey and top up with warm water.

A SPOONFUL OF VINEGAR

Just one spoonful of vinegar can make a difference to your health. It can be taken neat, or diluted in a glass of water.

WAXY EAR BUILD-UP Mix 15ml/1 tbsp each of white distilled vinegar and warm water and use as a gentle ear wash to remove any waxy build-up. Do not use this technique if you are suffering (or have suffered) from an ear infection or inflammation.

DIARRHOEA This is the body's natural way of eliminating toxins from the body. It can also be a symptom of a wide range of complaints. If your symptoms persist, seek professional advice. Cider vinegar will alleviate the intensity of diarrhoea but will not stop it completely. Take 5ml/1 tsp cider vinegar in a glass of water, approximately six times a day, before meals and in between.

IRRITABLE BOWEL SYNDROME Add 5ml/1 tsp of cider vinegar to a glass of water. Drink daily to relieve the symptoms.

CONSTIPATION Cider vinegar contains pectin, which can help your body maintain regular bowel movements. Take 5ml/1 tsp daily.

HICCUPS Try drinking a 5ml/1 tsp of cider vinegar to cure hiccups. If the attack of hiccups is particularly severe, try gargling with cider vinegar.

HEARTBURN This usually occurs 1–2 hours after eating. Drinking 5ml/1 tsp cider vinegar in a glass of water before a meal can help to prevent.

HEART HEALTH The potassium in cider vinegar helps to maintain a healthy heart and blood pressure.

Cholesterol

The grape juice and vinegar in this drink contain properties which may help to lower cholesterol levels.

INGREDIENTS

250ml/8fl oz/1 cup red grape juice

120ml/4fl oz/½ cup apple juice

20ml/4 tsp white distilled vinegar

1 Pour the red grape juice and apple juice into a large glass.

2 Add the white distilled vinegar to the glass, stir and chill for at least 1 hour in the refrigerator.

3 Drink a little of the mixture before main meals to help reduce cholesterol levels.

Lethargy

Cider vinegar, honey, mint and orange make an uplifting drink for anyone suffering from fatigue.

INGREDIENTS

475ml/16fl oz/2 cups water

6–8 fresh mint leaves

30ml/2 tbsp fresh orange juice

5ml/1 tsp cider vinegar

5ml/1 tsp honey

1 Bring the water to the boil in a small pan and add the fresh mint leaves.

2 Simmer for 2 minutes, remove from the heat, strain, and add the orange juice, cider vinegar and honey.

3 Serve warm or cool and serve chilled.

VINEGAR REMEDIES FOR COUGHS AND COLDS

Cider vinegar and honey are classic partners in the fight against everyday minor infections – soothing sore throats and blocked sinuses and assisting recovery.

Coughs and sore throat

Cider vinegar and honey both have natural antiseptic qualities so are excellent for relieving and soothing coughs and sore throats.

INGREDIENTS

30ml/2 tbsp cider vinegar

30ml/2 tbsp honey

1 Place the cider vinegar and honey in a small bowl and mix together well.

2 Take the occasional teaspoonful as needed to ease the irritation of a cough or sore throat.

Health tip

Try stirring the honey and vinegar mixture into a cup of hot water for a soothing drink.

VINEGAR GARGLE

Adding 15ml/1 tbsp cider vinegar to a cup of lukewarm water makes a good antiseptic gargle to help relieve a sore throat. Use this to complement cider vinegar and honey soothing drinks.

Blocked sinuses

A combination of cider vinegar, lemon juice and fresh mint leaves is both warming and clearing when a cold or blocked sinuses make breathing difficult.

INGREDIENTS

boiling water

juice of ½ lemon

small bunch fresh mint leaves

30ml/2 tbsp cider vinegar

1 Pour the boiling water into a large bowl, so that is is about half full.

2 Add the vinegar, lemon juice and fresh mint leaves.

3 Hold your head over the bowl of water, with a towel draped over your head to keep the steam in.

4 Inhale for 10–15 minutes, stopping if the steam becomes uncomfortably hot.

Health tip

Try adding other ingredients such as fruit or herb vinegars with eucalyptus and cardamoms.

Sore throat gargle

Use this soothing gargle at the first sign of a sore throat. It can also be taken internally, 10ml/2 tsp at a time, 2–3 times a day. Use within a week.

INGREDIENTS

small handful each of fresh sage and
    thyme leaves

600ml/1 pint/2½ cups boiling water

30ml/2 tbsp cider vinegar

10ml/2 tsp honey

5ml/1 tsp cayenne pepper

1 Roughly chop the fresh sage and thyme leaves and place them in a large jug (pitcher).

2 Pour the boiling water over the herbs, cover and leave for 30 minutes.

3 Strain off the leaves and stir in the cider vinegar, honey and cayenne.

4 Gargle with a mouthful of the mixture to soothe a sore throat. Keep refrigerated for up to a week, but allow it to warm up to room temperature before using.

VINEGAR REMEDIES FOR FIRST AID

A useful cleanser and cooler, diluted or neat vinegar can be helpful for its antiseptic properties. Keep a small bottle with your first aid kit.

Sunburn

A vinegar and water solution will help to soothe burnt skin.

To treat a case of sunburn, soak a clean cloth in equal parts water and white distilled vinegar. Cover sunburnt areas with the cloth before going to bed. Leave on overnight if possible. Alternatively, put the vinegar and water mixture into a spray bottle and spray directly on to burnt skin.

Burns

To treat minor burns, pour a little white distilled vinegar on to a paper towel and dab on to the affected area.

CAMPING TIP

Remember vinegar when picnicking or camping – in the absence of a formal first aid kit or antiseptic, vinegar is a useful cleanser for any minor scratches and cuts.

Cuts and grazes

Vinegar is a natural antiseptic. Making an antiseptic vinegar wipe for minor abrasions is less expensive and intrusive than harsh disinfectants.

Pour a small amount of cider vinegar on to a clean cotton wipe. Use to clean the skin around minor knocks and scrapes. It may sting a little, but it will help to prevent infection.

Hiccups

Drinking vinegar in water can halt an unfortunate attack of hiccups.

Add 5ml/1 tsp of cider vinegar to a glass of warm water. When suffering from hiccups, sip the mixture slowly and they will fade away.

Cold sores

Use vinegar to soothe the pain and swelling of cold sores.

Soak a cotton wool ball in white distilled vinegar. Apply to the affected area two or three times a day.

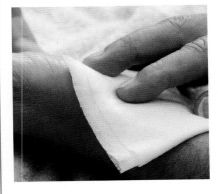

Bruises and swellings

Treat with a vinegar-soaked cloth.

Soak a cloth in equal parts white distilled vinegar and water. Wring out excess liquid. Apply to the bruise for at least 1 hour, holding in place with a bandage if necessary.

VINEGAR REMEDIES FOR CLEAR SKIN

A powerful astringent, a little cider vinegar diluted with spring water makes a zingy refresher or helpful potion to combat greasy skin.

Blocked pores

The alpha-hydroxy acids, or natural fruit acids, present in both strawberries and cider vinegar will act as a natural exfoliator when applied to the skin.

Acne

A gentle solution of cider vinegar and water can be used as an antiseptic face freshener.

Mix one part cider vinegar to 10 parts spring water. Transfer to a spray bottle, keep chilled and use as a face spritzer, making sure that the eyes are closed before spraying.

INGREDIENTS

3 large strawberries

55ml/¼ cup cider vinegar

1 Remove the stalks from the strawberries, and in a small bowl, mash them. Add the cider vinegar, mixing well. Allow the mixture to stand for 2 hours, then strain into a bowl.

2 Apply the strawberry and vinegar mixture to the face with cotton wool and leave on the skin for as long as possible, preferably overnight.

3 Wash the mixture off with cold water and a small amount of gentle cleanser if necessary. Your skin should feel soft and smooth.

Chapped skin

Soothe cracked or chapped skin with cider vinegar.

Mix equal quantities of cider vinegar and spring water. Apply gently to the skin with a cotton wool ball.

VINEGAR REMEDIES FOR FOOTCARE

Feet are subject to daily pounding and are vulnerable to skin irritations. A little vinegar in a cool footbath makes a refreshing, cleansing footwash.

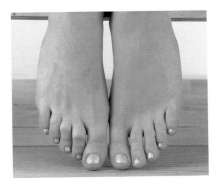

Dry or itchy skin

Soak in a vinegar bath to soften the skin on the feet, which is prone to dryness.

Add a few drops of cider vinegar to warm water. Soak the feet for 30 minutes to soothe dry or hard skin.

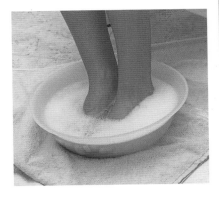

Corns

Traditionally, vinegar was applied as a bread poultice, by soaking bread in vinegar, then bandaging it on the corn. Now we know that corns need softening so that they can be gently rubbed off.

Soak the feet in warm soapy water and rub with olive oil to soften the corn. Rub off the corn with cotton wool. Use a dab of cider vinegar on a cotton wool ball to clean the area and prevent infection.

Athlete's foot

The cider vinegar in this foot bath helps to restore the pH balance of the skin, which becomes over-alkaline when suffering from this condition. Myrrh and tea-tree oil both have antifungal properties. If the condition is severe, with more than itching or with broken skin and blistering, it is important to follow medical advice. The ingredients can all be found at specialist health stores, or online.

INGREDIENTS

25g/1oz dried sage
25g/1oz dried pot marigold flowers
1 large aloe vera leaf, chopped
15ml/1 tbsp myrrh granules
2.2 litres/4 pints/9 cups water
10 drops tea tree essential oil
60 ml/4 tbsp cider vinegar

1 Pour the water into a pan and bring to the boil. Add the sage, dried pot marigold flowers, chopped aloe vera, and myrrh and cover. Allow to simmer for 20 minutes.

2 Allow the infusion to cool down a little, then strain into a large bowl.

3 Add the tea tree oil and vinegar to the bowl. Immerse the feet in the bowl for at least 15 minutes. Dry thoroughly.

TOENAIL FUNGUS

Drop a few drops of distilled white vinegar on the toenail three times a day to help clear up toenail fungus.

VINEGAR REMEDIES FOR HAIRCARE

Use vinegar in simple hair rinses, on its own or mixed with other ingredients, to promote general scalp health and to treat conditions such as dandruff, head lice and hair loss.

Dandruff rinse

Nasturtium is often used in hair products, and has a balancing effect on the skin. Use this hair rinse to soothe a dry, flaky scalp.

Hair loss

Vinegar is an excellent hair rinse for anyone and rinsing with cider vinegar may be helpful for those suffering from hair loss other than the genetic 'natural' balding experienced by men. An infusion of sage is also recommended – infuse sage leaves in a bottle of cider vinegar.

INGREDIENTS

25g/1oz nettle leaves

25g/1 oz nasturtium flowers and leaves

1 litre/1¾ pints/4 cups water

30ml/2 tbsp cider vinegar

30ml/2 tbsp witch hazel

1 Place the nettle leaves and nasturtium flowers and leaves in a large bowl.

2 Boil the water and pour it over the nettles and nasturtiums in the bowl. The nettle leaves will lose their sting in the boiling water.

3 Cover, allow to stand overnight, then strain off the leaves and flowers.

4 Add the cider vinegar and witch hazel to the strained liquid.

5 Pour through the hair as a final rinse every time you shampoo.

HEAD LICE (NITS)

To treat an attack of headlice, first apply warmed vinegar to the scalp. Cover the head with a shower cap for 30 minutes, then rinse. Dip a fine-toothed nit comb in vinegar and comb through the hair to remove the lice and their eggs. Repeat the treatment for 3–5 days, until the lice have disappeared.

INGREDIENTS

50g/1oz fresh sage leaves

bottle of cider vinegar, about 600ml/1 pint/ 2½ cups

1 Pour enough of the vinegar out of the bottle to allow room for the sage leaves.

2 Wash and drain the sage leaves, then place them in the bottle of cider vinegar.

3 Leave for 2 weeks to infuse, then use as a hair rinse after washing.

VINEGAR REMEDIES FOR INSECTS AND STINGS

As an emergency coolant, to cleanse and help reduce inflammation, and to keep them away, vinegar is useful for minor everyday insect attacks.

Insect bites

This will help to reduce any itching and irritation around the bite.

Soak a cotton wool ball in a small amount of white vinegar. Use this to clean the area around the insect bite.

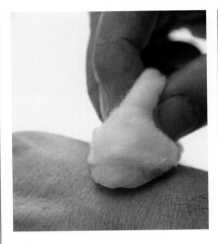

Jellyfish stings

Studies have shown that vinegar can deactivate the venom in jellyfish stings.

Dip a cotton wool ball in cider vinegar. Dab this on to the skin to reduce any pain and swelling. Seek medical attention.

Wasp stings

This will work for bee stings too but remove the sting from the skin first.

Soak a cotton wool ball in cider vinegar. Apply directly to stings to alleviate pain and swelling.

Insect repellent

Do not use a fruity sweetened vinegar as this may attract flies.

Pour a small amount of white vinegar into the palm of your hand. Rub the vinegar into the skin as a repellent.

Bug spray

This natural bug spray is pet- and child-friendly.

Transfer a small amount of white vinegar to a spray bottle using a funnel. Use it as an economical bug spray.

Mosquito repellent

Drink vinegar to keep mosquitoes away.

Drink 10ml/2 tsp cider vinegar a day to keep mosquitoes away. The vinegar is said to influence body odours – mosquitoes will find your perspiration unpleasant.

VINEGAR REMEDIES FOR MUSCULAR PROBLEMS

A refreshing vinegar bath soak will soothe tired or strained muscles, and will work wonders for the way you feel, aiding physical relaxation and calm.

Muscle ache

Tired muscles can be soothed by relaxing in a hot bath scented with a combination of cider vinegar and lavender oil.

INGREDIENTS

15ml/1 tbsp cider vinegar

2–3 drops lavender oil

1 Fill a bath with warm water.

2 Add the cider vinegar and the lavender oil to the water to create a soothing bath.

Muscle compress

Soothe sore or aching muscles with a warm vinegar compress.

1 Mix 475ml/16fl oz/2 cups each of water and cider vinegar together in a large bowl.

2 Soak a small towel or flannel in the vinegar mixture.

3 Heat the towel in the microwave for 20 seconds, until warm, then apply to the affected area.

MUSCLE TIREDNESS

A cider vinegar and honey drink is helpful for alleviating muscle tiredness after an exercise session. Put 5ml/1 tsp honey and 5ml/1 tsp cider vinegar in a cup and top up with hot water.

CRAMP

Vinegar is a traditional aid for relieving cramp, which can be caused by low levels of potassium. Take 5ml/1 tsp cider vinegar before going to bed to avoid waking up with cramp in the night.

VINEGAR REMEDIES FOR LATER LIFE

A little vinegar goes a long way to help combat some of the everyday symptoms of the long-standing conditions that can develop in later life.

Arthritis

Many people who suffer from arthritis swear by a daily dose of cider vinegar to help relieve the pain. Cherries are also commonly recommended to relieve the pain associated with arthritis and gout. Sufferers should make a batch of cherry cider vinegar to use as a dressing, in sauces or for drinks.

INGREDIENTS

450g/1 lb fresh or frozen cherries,
  stoned (pitted)

600ml/1 pint cider vinegar

1 Purée the fresh or frozen cherries in a blender with the cider vinegar.

2 Transfer to a bottle and store in the refrigerator. Shake well before using.

3 For a delicious drink, put 10ml/ 2 tsp cherry vinegar in a glass and top up with water.

Stiff joints

If the body is short of potassium, joints can become stiff. Taking a teaspoon of cider vinegar in a glass of water will help replenish potassium levels.

Denture cleaning

Soak dentures in a glass of vinegar for up to an hour, for a natural alternative to your usual denture cleanser.

Bone health

Cider vinegar contains the elements manganese, magnesium, silicon and calcium, which can help with bone health. Either take a cider vinegar drink daily, or try cider vinegar supplements.

Age spots

Mix together the juice of an onion with an equal amount of vinegar and apply daily to age spots. After a few weeks you will notice the spots diminishing.

NATURAL BEAUTY WITH VINEGAR

This chapter shows how many home-made vinegar-based natural beauty treatments can be used instead of commercial products, with fabulous results. It is a natural exfoliator and astringent, and is a valuable addition to many skincare preparations. From head to toe, whether in hair rinses, bath soaks, face packs, body scrubs or foot baths, learn how to use vinegar in many simple potions that are superbly effective and pleasingly inexpensive.

Left: Cider vinegar is a great addition to the bathroom cabinet, and features in many home beauty treatments.

A TRADITIONAL BEAUTY INGREDIENT

Some of the most expensive and exotic beauty products rely on pure natural ingredients that can easily be combined at home with fantastic results.

Vinegar has been used in beauty treatments for thousands of years. Helen of Troy – reportedly the most beautiful woman in the world – bathed in vinegar to tone and condition her skin.

More recently, kitchen ingredients were frequently used in the boudoir before modern cosmetics became relatively inexpensive. As well as vinegar, olive oil, honey, rose water, witch hazel, oatmeal and milk were all typical ingredients used for cleaning and conditioning the hair and skin. Even after bought potions became more common, the majority of women were limited to just one pot of face cream. Pampering sessions involved creating natural concoctions at home from tips and recipes which had been passed down the generations from mother to daughter and between friends.

Deliciously uncomplicated

Discovering home-made beauty potions is great fun! Mixing and applying simple natural ingredients to the skin can be very satisfying. Many modern beauty treatments can be extremely harsh on the skin (as well as being expensive). Many types of face scrubs and peels can be too fierce for frequent use. Also, using many commercial products can result in a build-up of some substances that have to be removed by applying another; this is especially true for hair products.

WHICH VINEGAR TO USE?

Cider vinegar has a pleasant, slightly fruity odour that makes a positive contribution to beauty remedies.

Organic apple cider vinegars benefit from the goodness of the apples used as the basic ingredient. Cider vinegar is the best choice for facial treatments.

Distilled white vinegar is perfectly suitable for the majority of the remedies described here – it has a light refreshing smell.

Fruit vinegars are a good choice for beauty as long as they are not sweet or sweetened. Sweet vinegars are completely unsuitable and should not be used.

Malt, balsamic and red or white wine vinegars are also not suitable for use in beauty treatments.

Above: Because cider vinegar is made from apples, it contains natural fruit acids, or alpha-hydroxy acids, which are very good for the skin, and are used in many cosmetics.

Above: Keep a bottle of cider vinegar handy on the bathroom shelf.

Above: Use a little vinegar as part of a natural manicure session.

Above: Oils, herbs, fruit, honey and oats have long been prized alongside vinegar for their natural cleansing and conditioning properties.

Above: Vinegar can be used alone or with herbs to make fantastic hair treatments.

Making vinegar treatments

The vinegar treatments suggested here are all easy and inexpensive to prepare. They will help to bring a natural glow of health to hair, skin and nails without causing any damage. Home-made treatments containing cider vinegar will exfoliate and brighten the skin, due to the natural fruit acids, or alpha-hydroxy acids, in the vinegar.

Many soaps and cleansers are very alkaline, and the mild acidic qualities in vinegar will help to restore the natural pH balance of the skin. Use home-made vinegar treatments to complement favourite moisturizers and similar products.

There are many uses for vinegar in the bathroom. Add it to bath water for a refreshing soak, or use it in body washes or cleansers. Vinegar is famous for its ability to neutralize odours, and can even be used as a natural deodorant. Use vinegar to remove stains from the skin, or to prolong the life of nail varnish. It is a fantastic product to use on the hair, and can be used in conditioning rinses on its own or with herbs and herbal oils to remove product build-up.

The natural antibacterial qualities of vinegar make it particularly suitable to use in treatments for the feet, which can be prone to fungal infections. Make vinegar foot scrubs and foot washes to leave your feet refreshed. Use the natural exfoliating and brightening properties of cider vinegar in body scrubs, toners and face masks to leave your skin soft and smooth from top to toe.

ALPHA-HYDROXY ACIDS AND VINEGAR

Natural fruit acids, known as alpha-hydroxy acids (AHA), act as exfoliators by removing dead skin cells and stimulating production of replacements and natural moisturizers. These acids are found in different proportions and mixes in different fruit, including apples, citrus fruit, papaya, strawberries, grapes and pineapple. They are also present in cider vinegar, making it a great choice when preparing natural beauty treatments.

VINEGAR FOR BATHING AND CLEANSING

Adding a little refreshing vinegar to a warm bath is brilliant for restoring tired bodies and will help to balance the skin's natural pH balance.

Bath bouquet

Using cider vinegar with herbs and orange will make a bath especially refreshing and relaxing.

INGREDIENTS

fresh mint, one bunch

fresh lavender, one bunch

rind of one orange

250ml/8fl oz/1 cup cider vinegar

baby oil

mild liquid bath soap

1 Tie the mint, lavender and orange rind together and place in the bath.

2 Pour the cup of cider vinegar into the bath on top of the bouquet.

3 Run a shallow layer of very hot water and leave to infuse for about 5 minutes.

4 Fill the bath with warm water to your preferred level, adding a few drops of baby oil and a mild liquid bath soap.

Head-clearing bath

Add a little cider vinegar, eucalyptus oil and tea tree oil to a hot bath for a head-clearing soak.

Scent neutralizer

To remove a scent that you have applied in error, or that has gone off, dab a little vinegar on to the skin that has been sprayed. The smell will be neutralized.

Deodorant

Using vinegar as a natural deodorant will not stop perspiration, but will neutralize body odour.

Cleanser

Most soap is alkaline, so using a mixture of equal parts vinegar to water helps to balance the skin's pH levels after bathing.

Exfoliating body wash

Vinegar is a natural exfoliator and this scrub is excellent for exfoliating arms and for encouraging circulation in the thighs and buttocks. It makes the skin all over smooth and moist. For a fabulous back scrub, enlist the help of a willing partner. Prepare a pad of towels and relax while someone gently scrubs and massages.

Bathroom aroma

A candle burner for heating aromatic oil in water is ideal for creating a relaxing environment for soaking in the bath. For a refreshing aroma, add a little white vinegar to the water – about 1.25ml/¼ teaspoon – as well as a few drops of oil.

INGREDIENTS

1 small handful rice, uncooked

5ml/1 tsp white vinegar

2.5ml/½ tsp olive oil

10ml/2 tsp gentle body wash or shampoo

1 Crush the rice in a mortar with a pestle so that it has a gritty texture. Place in a bowl and mix in the vinegar.

2 Add the oil to the bowl. Stir in the body wash or shampoo to make a creamy mixture.

3 Shower with hot water and then scrub with the mixture, lathering it up to a foam. Concentrate on knees and elbows if there are any signs of rough skin.

Refreshing bath

Pour some cider vinegar into the bath for a refreshing soak. This will also help to prevent mild bacterial and fungal infections.

Aftershave

A mixture of equal parts cider vinegar to water will work as an aftershave. The acetic acid in the vinegar acts as an antiseptic on cuts or abrasions.

VINEGAR FOR HAND TREATMENTS

A valuable cleanser and antiseptic, vinegar can be used to remove stains from the hands, to prepare fingernails for varnish and to prevent infection.

Longer lasting nail varnish

Apply vinegar underneath varnish to make it last longer before chipping.

Dip a cotton bud in vinegar. Use to clean the surface of the fingernail before applying a base coat or nail varnish.

Stain remover

This is a good way of removing fruit stains from the hands, for example after picking blackberries or strawberries. Margarine or lard and granulated (white) sugar make an excellent scouring agent for removing stubborn stains from hands when used with vinegar.

INGREDIENTS

45ml/2 tbsp vinegar

30ml/2 tbsp granulated (white) sugar

45ml/2 tbsp margarine or lard

a bowl of hot soapy water

vegetable oil

1 Prepare a small saucer each of vinegar, margarine or lard and granulated sugar.

2 Rub in the vinegar with a nail brush.

3 Scrub the hands with the fat and then finally the granulated sugar.

4 Repeat if necessary until the staining comes away, then wash in hot soapy water. Dry and massage with a little vegetable oil.

Hand health

When they are frequently in and out of water, hands are prone to fungal infection under and around the nail area. Use vinegar to prevent this.

Use a cotton wool ball dipped in vinegar to clean around and under the nail area. Dry under a hot air drier or shake the hands dry.

VINEGAR FOR HAIRCARE

Using vinegar in a hair rinse is a brilliant way to achieve super-shiny hair. It can also be used to lighten the hair or on dyed hair to keep colour strong and fresh.

Cider vinegar hair rinse

This gentle vinegar rinse will remove the build-up caused by using products such as hairsprays and waxes on the hair.

INGREDIENTS

washing up liquid

250ml/18 fl oz/1 cup cider vinegar

1 litre/1¾ pints/4 cups warm water

1 First use a little mild washing up liquid to shampoo the hair, massaging the scalp gently. Try to avoid rubbing the hair up into a tangle. Comb through any tangles with your fingers.

2 Repeat and then rinse the soap from the hair completely.

3 Put the cider vinegar into a suitable bottle, add the warm water and then use the mixture as the final hair rinse for shiny, healthy-looking hair.

Rosemary hair rinse

Use a vinegar rinse once a week to improve the health of your hair. Rosemary is a good conditioner for dark hair.

INGREDIENTS

50g/2oz sprigs of rosemary

250ml/18 fl oz/1 cup cider vinegar

10 drops rosemary oil

1 Infuse the rosemary sprigs in enough boiling water to cover them, for 12 hours.

2 Strain, add the vinegar and rosemary oil, then pour into a bottle to use as required.

CLEAN HAIR TONGS

If you notice a build-up of hair products on curling tongs or straightening irons, they can be cleaned with a vinegar solution. Mix equal parts water and vinegar with a pinch of salt.

Natural hair lightener

Lemon juice is known for its lightening effect on the hair. The acid in the vinegar will help to accelerate the process.

INGREDIENTS

juice of two lemons

120ml/4fl oz/½ cup cider vinegar

1 Mix the lemon juice and vinegar in a spray bottle. For subtle highlights, separate strands of hair and spray with the mixture, or spray hair all over.

2 Leave for 10 minutes then wash off.

COLOUR FIXER

When hair has been coloured with a natural product such as henna, use vinegar as a colour fixer. Once the henna has been washed out, apply 120ml/4fl oz/½ cup vinegar, massage into the hair and rinse.

VINEGAR FOR FOOTCARE

Inexpensive cider vinegar is ideal for everyday footcare – keep a small bottle in the bathroom cupboard and use these treatments regularly for softer skin.

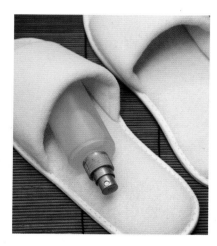

Refreshing spritzer

Make a vinegar spritzer to cool tired feet and legs.

INGREDIENTS

rosewater

cider vinegar

a few drops of eucalyptus oil

1 In a small, fine sprayer, mix 6 parts rosewater with 1 part vinegar and add the eucalyptus oil.

2 Shake well and place in the refrigerator to chill for at least an hour.

3 Spray lightly on tired feet and legs, then put the feet up while sipping a refreshing cider vinegar and herb or fruit tea.

TOENAIL CARE

Use a cotton bud dipped in vinegar to clean around toenails after trimming, filing, tidying and buffing. Then massage in oil, lotion or foot cream.

Rough skin scrub

Mix vinegar with crushed rice and oil to make a scrub which will leave your feet smooth and soft.

INGREDIENTS

1 handful rice, uncooked

5ml/1 tsp cider vinegar

2.5 ml/½ tsp oil

1 Crush the rice using a pestle and mortar to a gritty texture. Place in a bowl and mix in the vinegar until well combined, then add 2.5 ml/½ tsp oil. This makes enough for two foot scrubbing sessions. Keep the excess in a covered pot in the refrigerator for up to a week or freeze it.

2 Prepare a bowl of hot water for washing the feet after scrubbing and a small towel or clean dish towel on which to scrub them (this saves getting the scrub in the bottom of the bath).

3 Wash the feet in hot soapy water to soften the skin, then pat them dry. Place on the towel.

4 Take a spoonful of rice scrub in the palm of one hand and rub it into the foot, then use both hands to scrub the skin, working around the edges of the soles and toes as well as the heels to remove rough skin gently. Rinse well, then dry thoroughly.

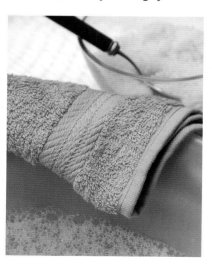

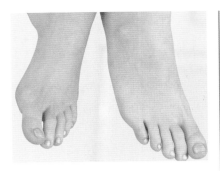

Soothing wax pack

This brightens and softens the skin as well as conditioning the nails.

INGREDIENTS

paraffin wax or nail or hand treatment wax, enough to cover both feet

vinegar, for massaging

2 pieces of clear film (plastic wrap), about 30x40cm

olive oil

1 Place each piece of clear film on a separate large folded towel.

2 Melt the paraffin wax or nail or hand treatment wax according to packet instructions in a large bowl over a pan of hot water. The bowl should be large enough for the toes to be placed in it.

3 Massage vinegar into one foot, working it well between the toes and around the nails. Check the temperature of the melted wax – it must be hot but not hot enough to be uncomfortable or burn the skin. Place as much of the foot as possible in the bowl and spoon the wax over it.

4 When the foot is coated in wax, place it on a piece of cling film. Spoon a little wax over any bare patches. Fold the cling film around the foot and then wrap the foot in the towel.

5 Repeat with the other foot. Relax for 30 minutes. Unwrap the feet and remove the wax. Massage with olive oil to finish.

Lavender and vinegar foot bath

Soothe worn-out feet with vinegar, relaxing lavender and lemon verbena.

INGREDIENTS

15g/½oz dried lemon verbena

30ml/2 tbsp dried lavender

5 drops lavender essential oil

30ml/2 tbsp cider vinegar

1 Put the lemon verbena and lavender in a basin and pour in enough hot water to cover the feet.

2 When it has cooled add the lavender oil and cider vinegar.

Footwash

Relax tired feet with vinegar and mint.

INGREDIENTS

12–15 mint leaves, crushed

3–4 drops tea tree oil

vinegar

1 Add the mint, tea tree oil and a little vinegar to a bowl of hot water.

2 Soak feet for 15 minutes.

FOOTCARE TIPS

• Washing in warm water with vinegar added is an old-fashioned treatment for all sorts of foot complaints, from corns to smelly feet. Vinegar deters fungal and bacterial activity, which is why it can help to prevent odours and foot infections. On its own it is not necessarily enough to cure these type of problems but it is a useful preventative lotion, for example to wipe out trainers and for cleansing feet immediately after using public swimming pools and showers.

• Walking barefoot in changing rooms is enough to pick up unwanted infection, so it is a good idea to wear flip-flops and wipe them over with a little vinegar when they are dry.

• As well as using vinegar in a footwash, a persistent problem of foot odour can be treated by rubbing a little vinegar over the feet after washing and drying every morning and night.

• Socks, tights or stockings should ideally be cotton and they should be washed daily. Remove them at the end of the working day, then wash and dry the feet before rubbing lightly with vinegar. Then leave the feet to air for the evening, if possible, rather than wearing socks or slippers.

• Keep footwear clean and allow it to air. Avoid plastics and materials that do not allow air to pass into the shoes and sweat to evaporate. Use suitable insoles that can be replaced or washed frequently. Sprays to prevent micro-organisms from infecting footwear can be expensive, especially for everyday use. Instead, wipe out shoes with a cotton pad dipped in vinegar and leave to air and dry every day.

• Try to avoid wearing the same pair of shoes too often. After sport, always remove training shoes from sports bags and allow them to air.

VINEGAR FOR SOFT SKIN

A cornucopia of natural oils, herbs and other gentle ingredients can be mixed with cider vinegar to make these fabulous skincare potions.

Refreshing hot cloth

This is a good pore-opening process that works very well before using a face scrub or applying a face pack. Eucalyptus oil has a lovely fresh smell.

Oatmeal salt scrub

Exfoliating vinegar is mixed with moisturizing oil, salt and oatmeal.

INGREDIENTS

30 ml/2 tsp fine oatmeal

30 ml/2 tsp salt

2.5 ml/½ tsp sweet almond oil or
   avocado oil

2.5 ml/½ tsp cider vinegar

1 Mix the fine oatmeal with the salt and sweet almond oil or avocado oil in a bowl. When thoroughly combined, stir in the cider vinegar.

3 Wash the face with hot water and mild soap, then pat dry, leaving the skin moist.

4 Gently exfoliate with the scrub, working around the nose and chin, up the cheeks and across the forehead. Avoid the delicate areas around the eyes.

5 Rinse well with hot water and pat dry.

INGREDIENTS

15 ml/1 tbsp cider vinegar

a few drops of eucalyptus oil

1 Add the cider vinegar and eucalyptus oil to a small bowl of boiling water. Place a clean face cloth in the bowl and swirl it around.

2 Remove any make-up and wash with hot water, then dry the face.

3 Drain the hot cloth and wring it out, leaving it moist but not wet. Take care not to burn the hands and be sure that the cloth is not too hot for the face.

4 Lay the cloth over the face, pressing it gently around the nose, chin and forehead. Lie back and relax for a few minutes while the cloth is hot.

5 When the cloth cools, remove it and apply a face pack or scrub while the pores are open.

Face steaming

This is a great way of opening the pores and drawing out dirt and oils. It is good before applying a deep cleansing mask, such as a mud mask.

INGREDIENTS

15 ml/1 tbsp cider vinegar
a few drops of lavender or tea tree oil
a sprig of fresh herb such as rosemary
 or mint
rind of 1 lemon or lime

1 Prepare a large bowl of boiling water and add the vinegar. Add your chosen essential oil and/or herbs.

2 Hold your face over the steaming bowl and drape a towel over both head and bowl. Your pores should be open and the face sweating profusely. Steam your face for 5–10 minutes, but stop if it becomes at all uncomfortable.

STEAMING TIPS

If you are unused to steaming your face take care not to have your face too close to the water. Lower your face slowly while pulling over the towel to make sure the steam is not too fierce. Electric face steamers are inexpensive and great for regular cleansing. Add a little cider vinegar to the water.

Rose petal toner

Infusions of flowers make excellent toners.

INGREDIENTS

40g/1½oz fresh rose petals
600ml/1 pint/2½ cups boiling water
15ml/1 tbsp cider vinegar

1 Put the rose petals in a bowl, pour over the boiling water and add the vinegar. Cover and leave to stand for 2 hours, then strain into a clean bottle.

2 Apply with cotton wool (cotton balls) after removing make-up. Keep chilled and used up within a few days as this will soon deteriorate.

Vinegar and rosewater freshener

Use this freshener around oily areas, such as the nose and chin. Avoid the area around the eyes.

INGREDIENTS

cider vinegar
rosewater

1 Mix 1 part cider vinegar with 3 parts rosewater and pour into a small bottle.

2 To use, dampen a cotton ball and then apply a little of the mix, then pat it over the skin to close the pores after cleansing.

Almond and raspberry face pack

This vinegar mixture will brighten skin.

INGREDIENTS

30ml/2 tbsp ground almonds

5ml/1 tsp unsweetened raspberry vinegar

Place the ground almonds in a bowl, add the unsweetened raspberry vinegar and mix well. Spread over the face using your fingertips. Leave for 30 minutes before rinsing.

Egg white face pack

The protein in egg whites can do wonders for the skin, particularly when mixed with cider vinegar and oil.

INGREDIENTS

1 egg white

5 ml/1 tsp cider vinegar

2.5 ml/¹⁄₂ tsp sweet almond or avocado oil

30 ml/2 tbsp cornflour (cornstarch)

1 Break up the egg white with a fork, then stir in the cider vinegar and sweet almond or avocado oil.

2 Add the cornflour (cornstarch) and stir it in well to give a creamy mixture.

3 Wash and dry your face, then spread the pack smoothly over the skin.

4 Lie down somewhere quiet and relax for about 30 minutes.

5 Rinse off the pack with warm water before moisturizing.

Brightening oat face pack

Combine moisturizing oil with vinegar.

INGREDIENTS

30ml/2 tbsp rolled oats

5ml/1 tsp cider vinegar

5ml/1 tsp olive oil

3 drops of lavender oil

Mix the oats with the cider vinegar and olive oil. Add the lavender oil and spread over the face, avoiding the eyes. Rinse after 30 minutes.

Cucumber and mint face pack

Use the natural cooling and cleansing properties of cucumber with cider vinegar to make a refreshing face pack.

INGREDIENTS
cucumber, about 2.5cm/1 in length
fresh mint leaves
2.5ml/½ tsp cider vinegar
15ml/1 tbsp fine oatmeal

1 Wash and finely grate the cucumber (skin and all) into a small bowl.

2 Add a few chopped fresh mint leaves and mix in the cider vinegar. Stir in the fine oatmeal and spread the mixture over the face.

3 Chill out for 15 – 30 minutes before rinsing with lukewarm water and spritzing with chilled spring water.

Vinegar, parsley and yogurt pack

Parsley and cider vinegar are both known for their astringent qualities. This soothing mask is great for oily skin.

INGREDIENTS
a handful of fresh parsley
5ml/1 tsp cider vinegar
15ml/1 tbsp rolled oats
30ml/2 tbsp low-fat yogurt

1 Finely chop the fresh parsley and place it in a small bowl. Add the vinegar and mix well.

2 Stir in the rolled oats until thoroughly combined, then gently mix in the low-fat yogurt.

3 Spread this all over the face, avoiding the eyes. Relax for 15 minutes before washing it off.

ENLARGED PORES

Mix 30ml/2 tbsp almond flour with enough water to make a paste. Apply evenly to the face, paying particular attention to areas with enlarged pores. Leave for for 20 minutes. Rinse with warm water, then apply a solution of apple cider vinegar and water.

Stimulating fruit paste

Papaya contains alpha-hydroxy acids, (also present in the vinegar), which have an exfoliating effect on the skin.

INGREDIENTS
fresh papaya
15ml/1 tbsp cider vinegar
15ml/1 tbsp plain low-fat yogurt

1 Peel and deseed the papaya. Cut two slices about 1cm/½ inch thick.

2 Dice the flesh of the two papaya slices and place in a food processor or blender.

3 Add the cider vinegar and yogurt and blend to make a smooth paste.

4 Cleanse the skin as usual, then apply the fruit paste evenly to the face, taking care to avoid the delicate skin around the eyes.

5 Leave for about 20 minutes, then rinse off with warm water.

Beauty tip

If papaya is difficult to get hold of, you can use other fresh fruits such as pineapple, peach, strawberries or grapes.

VINEGAR IN THE HOME AND GARDEN

Walls and windows, floors and furniture can all sparkle with the aid of a little vinegar. Inside and outside, this chapter includes all sorts of suggestions for effective vinegar alternatives to costly commercial cleaning products, from polishing mirrors, cleaning paintwork, and removing stains to deodorizing rooms, trapping insects and killing weeds. There are ideas for laundry, car valeting, home decorating and gardening, as well as using vinegar to help ensure pets are in peak condition.

Left: Once you start using vinegar instead of chemical cleaning products you will be amazed at its effectiveness.

VINEGAR AROUND THE HOME

Vinegar brings a sparkle as well as a fresh aroma to many everyday cleaning tasks, from refreshing the refrigerator to making windows gleam.

Vinegar has a great variety of traditional uses as a household cleaner. It can be used for minor everyday tasks as well as renovation and restoration projects. It is a traditional stain remover for laundry as well as a type of 'rinse aid' for shiny surfaces that are to be polished off with a clean dry cloth. It is also useful for removing built-up polishes and grime found on old wooden furniture and fittings.

Discovering vinegar as a cleaner and restorer is exciting because it is neither as noxious nor expensive as many of the harsh commercial products can be.

As many household uses for vinegar were established generations before modern fabrics and furnishings were even invented, many of the methods are particularly useful for treating old and antique items, such as furniture or even old linens that can easily be damaged by modern detergents or bleaches.

Cleaning power

Vinegar is also useful for cleaning modern materials and surfaces. It is less harsh and less expensive than the majority of cleaning products and ideal for everyday use, especially to prevent a build-up of dirt, such as scale or grease in sinks, wash basins and drains.

Vinegar is great for cleaning glass, leaving it sparkling and free from greasy smearing. As well as windows, mirrors and items of glassware, try vinegar for glass splash-backs, doors and surfaces in the kitchen and bathroom. It is also excellent for polishing many modern

SCENTED VINEGAR

Add lavender or mint sprigs and strips of lemon rind to a bottle of distilled malt vinegar that is to be used for cleaning to give a fresh aroma.

metal finishes, including stainless steel and chrome, shiny rigid plastic surfaces and paintwork. It can be used to clean appliances such as irons, dishwashers, kettles and coffee-makers.

Vinegar may be used in the first stage of cleaning, for example to wipe off greasy stains from a surface, or as a final polish after washing down with hot soapy water. Vinegar can be added to the final rinsing water after washing. Alternatively, when the surface is washed and completely dried, vinegar can be applied as a polish by using it on a clean lint-free cloth. Vinegar can be used in the garden for cleaning or restoring furniture, as well as an insect and cat repellent.

General preparation

It is easy to forget that vinegar is an acid, and can damage as well as clean, so before launching in with bottle and cloth or brush, take a few moments to check

Above: Vinegar is particularly suitable for use on antique wooden furniture, but it should always be tested on a small out-of-sight area first.

WHICH VINEGAR TO USE?

Malt vinegar is the most economical choice for household and garden use as it can be purchased in large quantities and is less expensive than wine or cider vinegars. Distilled white vinegar is clear and therefore better than dark vinegar, which may stain fabrics. However, when cleaning extremely dirty old wood or tarnished metal, ordinary malt vinegar does just as well, especially in the first stages.

Similarly, for garden and outdoor use, inexpensive malt vinegar is the most sensible choice, if there is no danger of staining. Before cleaning new items, fabrics or surfaces with vinegar, always read the manufacturer's instructions to check that using vinegar (a mild acid) will not cause any damage.

the effect vinegar may have on the item and to protect the surrounding area. Begin by reading the manufacturer's instructions for a particular item.

Always experiment on a small area of fabric, furniture or material that is hidden or, preferably, separate from the item to be treated to make sure that the solution used will not cause any adverse reaction or damage.

Ensure that surrounding areas are covered or protected in some way, or that the items to be cleaned or restored are removed. For example, door fittings, such as handles, should be removed if they are to be restored to avoid damaging surrounding wood. Paintwork or frames around mirrors or glass should be protected by applying a suitable masking tape; conversely, when cleaning frames, it is important to avoid damaging mirrors, glass or the contents of a frame.

Gather together everything required before starting work. This is particularly practical for cleaning, rinsing, drying and polishing with efficiency when speed is important.

Above: Storing a bottle of vinegar in the cupboard with other cleaning products and materials will help you get into the habit of using it around the house.

Above: A mixture of vinegar and fruit juice is used to attract flies away from grapes.

Above: Distilled white vinegar can be used to clean resilient surfaces.

VINEGAR FOR FURNITURE, FRAMES AND MIRRORS

Vinegar has a long history of use by professional polishers and restorers on wooden furniture — it is good for gentle renovation rather than harsh stripping.

Wooden furniture

Vinegar and oils are traditional cleaning and restoring agents for old wood. A paint brush, toothbrush or special curved furniture brush are useful for getting into awkward corners or cleaning moulding and joinery.

INGREDIENTS
60ml/4tbsp vinegar

wax polish or furniture oil

1 Pour the vinegar into a dish and dampen a brush or small piece of wire wool, then rub a small hidden area of the item.

2 Wipe with a damp cloth and then with a dry cloth. Check the result — depending on the finish and the extent of cleaning required, a quick rub and polish will remove a light build-up of old polish and dirt.

3 When a small patch has been cleaned successfully, begin to work all over the item. Clean small areas at a time, being consistent, and drying them well. Change the vinegar and wire wool frequently.

FURNITURE TIPS

Wear rubber gloves to avoid staining the hands and nails. Always work with the grain of the wood, rubbing along it, not against it. Use light pressure and repeat the process rather than being too fierce with one application.

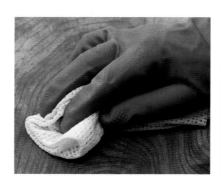

4 When the item is clean, give it a rub over with a damp cloth moistened with vinegar. Quickly dry it before buffing to a polish with a suitable oil or wax polish.

Mahogany bloom

Polished mahogany furniture sometimes develops a bloom or 'misty' surface, which can be prevented with vinegar.

INGREDIENTS
15ml/1 tbsp vinegar

5ml/1 tsp linseed oil

5ml/1 tsp turpentine

1 Prepare two solutions: in one bowl add the vinegar to 120ml/4 fl oz/½ cup hot water. In another bowl, mix the linseed oil and turpentine with 600 ml/1 pint/2½ cups water.

2 Wring out a cloth in the hot water and vinegar and wipe the wood. Follow with a cloth wrung out in the oil and turpentine solution. Polish with a chamois leather to dry the wood and give it a shine.

3 Repeat if necessary. Wipe down with vinegar water occasionally to prevent a build-up of bloom.

REMOVING RINGS

If wet drinks glasses have left rings on wooden furniture, remove them by rubbing with a mixture of one part olive oil, one part white distilled vinegar.

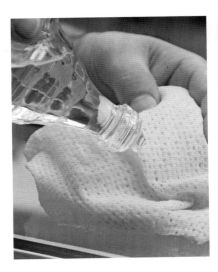

Glass tabletops

Rub glass surfaces with a vinegar cloth to remove any smears and greasy marks.

Wash the tabletop or surface as usual with soapy water and dry with a clean dish towel. Then dampen a clean cloth and sprinkle a little vinegar over it. Rub the surface with the vinegar cloth then polish it again with a clean dry cloth.

Lacquer

Bowls and ornaments will polish well if vinegar is added to the washing water.

Add 15ml/1 tbsp vinegar to a small bowl of warm water. Place the item in the mixture. If the item cannot be submerged in water, then wipe it with a cloth dampened with the solution. Dry well with a clean dish towel and then polish with a clean dry cloth.

Gilt frames

To clean a gilt frame use vinegar with a very soft brush that will not scratch it.

Dust the frame, then clean with a very soft brush dampened with a mixture of 1 part vinegar to 4 parts water. Rinse the brush frequently to remove the dirt and change the water when it is dirty. Gently polish the gilt with a clean dry cloth when it is dry.

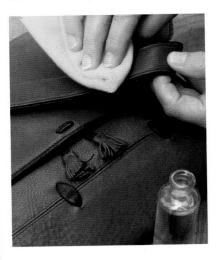

Mirrors

This is ideal for removing soapy splashes from bathroom mirrors and mirror doors.

Hold a cloth under running water, then wring out any excess water. Sprinkle a little vinegar on to the cloth and use to clean the mirror. Dry and polish with a clean cloth.

Light marble

Use a little white vinegar to remove stains from light marble.

Brush vinegar on the stained area of the marble and leave for at least 30 minutes to bleach out. Wipe the vinegar off the marble with a damp cloth.

Leather

To clean, revive and preserve leather items, use vinegar and boiled linseed oil.

Mix 1 part vinegar to 3 parts linseed oil in a small bowl. Rub the mixture into the leather with a soft cloth. Polish with a clean soft duster.

VINEGAR FOR WINDOWS, WALLS AND FLOORS

Known for its ability to cut through grease, vinegar gives a bright finish to glass, mirrors and glossy paint finishes, as well as being a very effective cleaner for carpets and tiles.

Windows

This old-fashioned method for cleaning glass is well known and very effective. Using vinegar and newspaper will ensure windows are smear-free.

INGREDIENTS

600ml/1 pint/2½ cups vinegar

1 Soak a cloth in the vinegar.

2 Clean the windows with the vinegar-soaked cloth.

3 Rub down the windows with scrunched up sheets of newspaper.

4 If the outsides of the windows are particularly dirty, then wash well first with warm water and finish with vinegar and paper.

GREENHOUSES

This method is also excellent for greenhouses. As a simple alternative, wipe over the windows with a vinegar cloth and then polish them with a clean, dry lint-free cloth.

Chewing gum

Use vinegar to remove hardened chewing gum from clothing, furniture or carpets.

Test a little vinegar on a hidden area first. When applying the vinegar to fabric, soak the whole area around the gum. Warming the vinegar for a few minutes in a small pan first helps when removing gum from furniture, carpets or flooring.

Window films

These can be very difficult to remove. Take care to wipe this vinegar solution off surrounding frames to avoid damaging paintwork or varnish.

INGREDIENTS

600ml/1 pint/2½ cups vinegar
45ml/3 tbsp washing soda

1 Mix the vinegar with 600ml/1 pint/ 2½ cups boiling water in a bowl. Stir in the washing soda. Wearing rubber gloves, wash the glass with this solution. It will soften the glue so it can be rubbed off.

2 A razor blade or glass cleaning blade designed for removing paint from windows can be used to scrape off any tough bits of plastic or glue residue.

Gloss paintwork

To clean gloss paintwork, use vinegar mixed with water.

Add 250ml/8fl oz/1 cup vinegar to a large bowl of warm water and wipe the mixture over the paintwork with a damp cloth. Dry and polish the surface with a lint-free cloth. To bring a gleam to large areas of paintwork, dust and wipe down as usual, then dry with a clean cloth. Sprinkle a little vinegar on to a clean damp cloth and rub down the paintwork, then polish with a clean cloth.

Floor rugs

Use a little vinegar in warm water to clean and restore floor rugs.

Vacuum clean, shake and/or beat the rug first to remove as much dirt and dust as possible. Brush it thoroughly. Add 250ml/8fl oz/1 cup of vinegar to a bucket of warm water. Test a little of the solution on the underneath to make sure the colours do not run. Then dampen a handbrush and brush the rug lightly all over with the solution. This will brighten colours and remove dirt.

Masking tape residue

If masking tape is left on windows after decorating the frames, the glue deposit left behind can be difficult to remove.

If there is a hard, yellowed deposit that can be scraped off using a blade, do this first – use a sharp blade at an acute angle to avoid scratching the glass. Mix 600ml/1 pint/ 2½ cups vinegar and 600ml/1 pint/2½ cups boiling water in a large bowl. Stir in 45ml/ 3 tbsp washing soda. Rub on the vinegar and soda solution. Wash and repeat until all the deposit has been removed.

Carpet stains

Use a vinegar and salt mixture to remove light carpet stains. Test on an out-of-sight patch before use.

In a small bowl, mix 30ml/2 tbsp salt with 110ml/½ cup vinegar until the salt has all dissolved. Rub the mixture into stains, allow to dry, then vacuum.

Wall tiles

Clean tiles with a damp cloth sprinkled with vinegar then polish with a dry cloth.

For thorough cleaning, for example to remove cooking splashes, rub with vinegar, then wash off with soapy water. When dry, polish with a damp cloth and a little vinegar, followed by a dry cloth.

Floor tiles

Vinegar is an excellent rinse aid for the final water used after cleaning a floor.

Clean floor tiles as usual. Add 250ml/ 8fl oz/1 cup vinegar to a bucket of warm water. Use the vinegar and water mixture to give the floor a final rinse to leave it gleaming.

VINEGAR FOR FIXTURES AND FITTINGS

White distilled malt vinegar is ideal for polishing gleaming fittings but ordinary dark malt vinegar is fine for restoring older grimy metals.

Old handles

Metal furniture handles that are old and have become tarnished and filthy can be removed and cleaned thoroughly with vinegar.

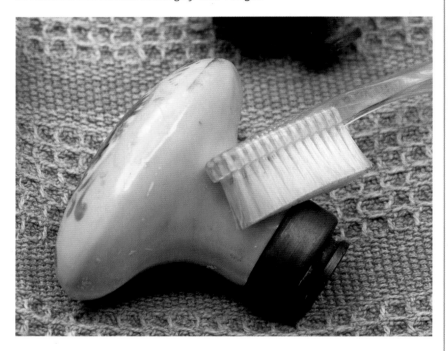

Black grates

Use vinegar to polish black fire grates or stoves.

Use a cloth dampened with vinegar for rubbing on the blacklead polish. This will remove grease and dirt and improve the quality of the polish.

INGREDIENTS

60ml/4 tbsp vinegar

1 If the item is encrusted and very tarnished, it may need soaking briefly in vinegar. Brush off any loose dirt, then place the item in a saucer or shallow dish and pour the vinegar over it.

2 Brush the surface with a small paint brush. Remove the handle from the vinegar as soon as the metal begins to look cleaner.

Variation

Particularly dirty items can be boiled. Place in a small pan and cover with vinegar, then bring slowly to the boil. A few minutes' simmering is usually enough but the item may have to be soaked in the hot vinegar, in which case remove from the heat, cover and leave to stand. Always clean with a soft brush that will not scratch the metal, then rinse the item in hot soapy water, dry thoroughly and polish with a clean cloth.

Venetian blinds

This technique works wonders on grimy Venetian or slatted blinds.

Mix together vinegar and warm water in equal parts. Briefly soak the fingers of a white cotton glove in the mixture. Put on the glove and slide your fingers along the slats of the blind, top and bottom. The dirt will come off in your hands.

Brass fittings

This old-fashioned method was used for cleaning chandeliers or similar fine brass fittings.

Heat 60ml/4 tbsp vinegar. Fold a cloth into a pad, dip it in the vinegar and sprinkle with salt, then rub this over and around the fitting. Rinse with warm soapy water and dry thoroughly. Polish the metal with paraffin oil or a light oil, such as sweet almond oil.

Rust

Use vinegar with salt to scrub off small patches of rust on intricate metal fittings.

Prepare two small bowls, one of vinegar, about 60ml/4 tbsp, and one of salt. Dip a stiff toothbrush in the vinegar, then in the salt and use to scrub off rust. Rinse the toothbrush in the vinegar then use to brush off the salt. Dry the metal and buff it with a clean lint-free cloth.

Showers

Vinegar is a great cleaner for shower heads, curtains, doors and fittings.

Clean the shower head with a solution of equal parts vinegar and water to remove deposits. Polish fittings with the same mixture and use to wipe over the tiles before buffing. Polish glass shower doors with a damp cloth sprinkled with vinegar. When washing out a shower curtain, add 50ml/ 2 fl oz/¼ cup vinegar to the final rinse.

Stainless steel fittings

Polish stainless steel fittings with vinegar for a gleaming finish.

Soak a sponge in hot water and sprinkle with a little vinegar. Squeeze most of the water out of the sponge. Use the sponge to rub stainless steel handles and fittings.

Chrome

Polish chrome items and fittings with vinegar and water.

Soak a cloth in a solution of equal parts of vinegar and water. Wring out the cloth and use to polish chrome fittings. Buff with a clean dry cloth.

Taps

Use vinegar to remove the dirt and limescale that can build up on taps.

Soak a paper towel in vinegar for 5 minutes. Wrap around the tap. After 1 hour, remove the paper towel and clean as usual. This will remove limescale and hard water deposits.

VINEGAR FOR REMOVING ODOURS

For wiping or spraying, vinegar is a useful deodorizing agent for combating unpleasant odours around the home, from smoky fireplaces to strongly-smelling foods in the kitchen.

Room odours

Adding a little vinegar to hot cinders will remove the smell of smoke from a room with an open fire.

INGREDIENTS

hot (but nor glowing) cinders or charcoal

120ml/4fl oz/½ cup vinegar

1 Shovel the hot cinders into a suitable metal bucket or empty coal scuttle.

2 Place this on the hearth or other heatproof stand. Pour on the vinegar.

Variation

If the smoky smell has spread to rooms with no fireplace, heat some barbecue charcoal in a large roasting tin in the oven on the hottest setting until it is very hot. Place this on a suitable heatproof base in the room and pour over the vinegar. As the cinders or charcoal cool, any smoky smells will be neutralized.

Smoky rooms

The smell of smoke from cigarettes can often linger on soft furnishings for several days. With this simple method, all traces of the smell should be gone in less than 24 hours.

Fill a small, shallow bowl about three-quarters full of cider vinegar. Place the bowl in the room where the smell is strongest. If necessary you can use several bowls in different rooms to get rid of persistent odours.

DRAIN DISORDERS

Try using a little vinegar as a first-stage solution for dealing with slightly smelly drains. Pour 250ml/8fl oz/1 cup vinegar into the problematic drain and leave it for several hours. Flush out the drain with very hot, soapy water. Repeat the process if necessary.

Food smells

For a gentle and natural alternative to aerosol air-fresheners, try this vinegar mixture. It is particularly effective for removing strong cooking smells in the kitchen.

Mix together 5ml/1 tsp baking soda, 15ml/1 tbsp distilled white vinegar with 475ml/16fl oz/2 cups water. Transfer the mixture into a small spray bottle, and spray into rooms as required to remove odours.

GARBAGE DISPOSAL UNITS

Bits of food can get caught in the blades of garbage disposal units, causing unpleasant odours. To prevent this, make vinegar ice cubes with a mixture of one part vinegar and one part water. Run the ice cubes through the garbage disposal unit every few weeks to keep it clean and odour-free.

Refrigerator hygiene

The refrigerator should be cleaned regularly with a little vinegar.

Remove all the shelves and clean with hot soapy water. Give the interior a final wipe over with a damp cloth wrung out in hot water to which a little vinegar has been added. This will remove any smells and leave it fresh and clean.

Cooking smells

Heat a small amount of vinegar on the hob to reduce food smells.

Pour a 120ml/4fl oz/½ cup vinegar into a small pan. Bring to the boil then place over a low heat, for the duration that the food giving off the smell will be cooking.

Musty cupboards

Wash cupboards and shelves out with this solution.

Add 225ml/1 cup vinegar, 225ml/1 cup ammonia and 60g/¼ cup baking soda to 4 litres/3 pints warm water. Dampen a cloth in the mixture then use to wipe down the interior of the cupboard. Leave the doors open until dry.

Toilets

Use vinegar to deodorize and freshen up the bathroom.

To freshen up the toilet, pour 600ml/ 1 pint/2½ cups vinegar into the bowl. Leave for at least 30 minutes before flushing the vinegar away.

Smelly ovens

Use a mixture of equal parts vinegar and water to remove lingering odours from ovens.

Place a 120ml/4fl oz/½ cup each of vinegar and water in a suitable heatproof container in the oven and heat until it is boiling. Turn off the heat and leave the vinegar water to stand in the oven overnight. The next day, wipe the oven out with a damp cloth which has been wrung out in vinegar and water. The heated vinegar will remove odours, as well as softening any hard, dried-on food, making it easy to wipe clean.

Microwaves

Use a little vinegar to clean a microwave.

Heat equal amounts of vinegar and water in the microwave for 2 minutes, then wipe down the insides with a damp cloth.

Oven-cleaner smells

Chemical oven cleaners can have a very strong smell which may linger for several days after cleaning.

Use the oven cleaner as directed on the packaging. To neutralize any strong smells, wipe down the oven with vinegar directly after using the chemical cleaner.

VINEGAR FOR THE KITCHEN

Vinegar is an inexpensive alternative to commercial cleaning products, and can be used in a variety of ways, from cleaning cupboards to removing stains from vases and chopping boards.

Biscuit tins and bread bins
Empty the crumbs and clean out biscuit tins and bread tins regularly.

Wipe down tins with a cloth sprinkled with neat distilled white vinegar. Leave tins open so that they dry out completely before using. This method is also good for lunch boxes.

Thermos flask
Remove tainting from strong flavours.

Crush the shell of one egg and add to the flask with 15ml/1 tbsp vinegar. Cover with warm water and seal. Shake well and empty into a bowl. Repeat if necessary, then rinse well.

Stainless steel pans
Vinegar will leave pans gleaming.

Pour 30ml/2tbsp vinegar into a stainless steel pan and clean with a brush. Rinse with hot water. Repeat if necessary. For the outside of the pan, use vinegar on a cloth with a little salt as a scouring agent.

Kitchen cupboards
Wipe shelves with vinegar and water.

Fill a spray bottle with half and half distilled white vinegar and water, then use for wiping down the insides of cupboards. Leave the cupboards empty and open to dry if time allows.

Pastry brushes
Vinegar is useful for cleaning the oil from pastry brushes.

Soak the pastry brushes overnight in a glass of vinegar. Add a little washing up liquid and work out the oil, then wash in hot, soapy water.

Copper pans and utensils
Bring back the sparkle to copper.

Wash the items in hot water. Use a mixture of sand and salt as a scouring agent and mix in a little vinegar to moisten. Rub over the copper with a cloth, adding more vinegar as required. Rinse with warm water.

Stained vases

This is an old-fashioned cleaning method using vinegar and tea leaves. It is good for removing stains from the insides of vases.

Pour warm water into the stained vase so that it is about two-thirds full and add 15ml/1 tbsp each of vinegar and tea leaves. Shake well and leave the mixture to stand for at least three hours. Shake the mixture occasionally, checking on the progress of the stain removal. Drain the vase and rinse it out well with clean warm water.

Stained mugs and cups

The insides of cups and mugs often become unpleasantly stained with tea and coffee, especially when they are not rinsed and washed immediately after use.

Pour in 15ml/1 tbsp vinegar and scrub with a washing up brush, then wash and dry. For really stubborn stains, especially on glazed pottery, scrub with a little vinegar and leave to stand for a while before rinsing. Alternatively, add a little vinegar to hot water in the mug and leave to soak for a few hours.

Burnt pans

Use vinegar to remove burnt-on food from the inside of pans.

If food has burnt slightly on the bottom of a pan, add 30ml/2 tbsp vinegar and a little water, to cover the bottom of the pan, then bring to the boil. Remove the pan from the heat and cover, then leave to soak to soften the burnt-on food. If the burning is fairly minor, then heating the pan with a little vinegar, allowing to cool and scrubbing is usually enough. For more severe burning, the pan may be left to soak overnight.

Glass carafes and decanters

This is a simple and effective approach to cleaning glass items.

Pour a little vinegar into the container, swill it around and add warm water. Leave to stand for a few hours or overnight. This is particularly useful for lightly stained items.

Chopping boards

Remove germs and staining from boards.

Wash boards well in hot soapy water. Drain and scrub with distilled white vinegar, sprinkling with salt for scouring, if needed. Rinse under hot water and allow to drain and dry before storing.

Cutlery

Clean tea or coffee stains on spoons, or water stains on stainless steel cutlery.

Dip cutlery in a little vinegar and rub with a cloth or kitchen paper. Wash in very hot water and dry immediately, polishing with a dish towel.

VINEGAR FOR LAUNDRY

Using vinegar is an old-fashioned and effective way of preventing the colour from running in bright fabrics. It has several other laundry uses too.

White fabric
Use vinegar to whiten yellowed fabric.

Wash and rinse the fabric. Rinse out in a little vinegar and water, and hang out in bright sunlight to dry.

Scorched fabric
This will remove scorch marks caused by ironing cotton or linen. Fuller's earth clay is available from craft shops.

INGREDIENTS
30ml/2 tbsp soap trimmings or soap flakes
2 large onions, peeled and grated
20ml/4 tbsp fuller's earth clay
250ml/8fl oz/1 cup vinegar

1 Place the soap trimmings or soap flakes in a pan and add 250ml/8fl oz/ 1 cup of water.

2 Heat gently, stirring, until the soap has completely dissolved.

3 Place the grated onions in a sieve (strainer) over the pan. Press out the juice from the onion into the dissolved soap.

4 Add the fuller's earth clay and the vinegar to the soap and onion mixture and stir well. Bring to the boil, then remove from the heat, stir well and cover. Leave to cool.

5 Spread out the scorched area of fabric, laying it over a pad of cloth or on a plastic rack. Spread the cooled mixture over the scorch mark and leave it to dry completely. Then wash out the fabric and let it dry in bright sunlight.

Laundry baskets
This is particularly useful for cleaning basketwork items.

Wipe down the inside of laundry baskets with a damp cloth sprinkled with a little vinegar. Leave to dry completely before using.

Dishcloths, brushes and sponges

If a dishcloth is to be left unused for some time, or after clearing up at the end of the day, use vinegar for a thorough clean.

Rinse out cloths in a bowl of water with a little vinegar added. Open out and hang up to dry. Rinse brushes and sponges in water with vinegar added, shake off or wring out and leave to dry.

Terry nappies (diapers)

Place dirty terry nappies in a bucket of water and vinegar to soak before washing. This will neutralize odours.

Add 300ml/½ pint/1¼ cups distilled white vinegar to a 5-litre/1-gallon bucket water, and leave the nappies to soak until it is time for them to be washed. When it is time for the nappies to be washed, drain the vinegar water, rinse and wash as usual.

Coloured fabric

Just as vinegar restores colour to red cabbage that has been steeped in salt before pickling, vinegar can also be used to brighten up washed-out fabric.

Wash clothing as usual. Add 250ml/8 fl oz/ 1 cup vinegar to rinsing water to restore colour. The acid in the vinegar is mild enough so that it will not harm clothing, but strong enough to dissolve alkalines in detergents.

Swimwear

Use vinegar to clean swimwear and remove the chlorine smell that comes from swimming in public pools.

Wash out swimwear in soapy water. Rinse in water with a little vinegar added – this will remove any traces of chlorine.

Wine stains

This works best if done within 24 hours of staining.

Sponge neat white distilled vinegar on to the affected areas and rub away the stains. Wash according to the garment's care instructions.

Fabric shine

After ironing, dark fabrics can develop a shine. White vinegar will remove this.

Transfer a little vinegar to a spray bottle. Spray a little on to the affected area, then sponge off with cold water. Test a small hidden area of fabric first.

VINEGAR FOR DECORATING

Grease-free surfaces are vital for good results when decorating and vinegar is the ideal cleansing agent during the preparation stages.

Woodwork preparation

When freshening up woodwork with a coat of paint, use vinegar to clean off grease and dirt first.

Pour a little vinegar on to a cloth and wipe down all woodwork which is to be painted. Then wipe with water, using a damp cloth, and dry thoroughly with paper towels. Sand down the surface ready for the new paint.

Wallpaper

If washing soda is used to strip wallpaper, use a vinegar solution to neutralize it so that it does not soak in and discolour new layers of paintwork.

To neutralize the soda solution, wipe down the walls with a vinegar solution, one part vinegar to two parts water. Leave to dry completely before sanding down and continuing with the painting.

Paint brushes

Depending on the type of paint and severity of the build-up, vinegar can be an effective relatively mild cleaner.

INGREDIENTS
250ml/8fl oz/1 cup vinegar
washing soda

1 Soak the paint brushes up to the tops of their bristles in the vinegar for several hours or overnight, then wash out with detergent and hot soapy water.

2 If the paint build-up is severe, the best way of cleaning brushes is by soaking them in a strong solution of washing soda overnight. This will release dried paint and soften the top of the bristle area.

3 Wear rubber gloves and use detergent to lather out all the paint into the soda solution. Rinse the brush in hot water.

4 Finish by soaking in a vinegar and water solution to neutralize any remaining soda. Rinse thoroughly, shake out and leave to dry.

DECORATING ODOURS

Place small dishes of vinegar around a room while decorating, to reduce the potency of odours from paint or varnish.

VINEGAR FOR PET CARE

Always check out traditional home cures for pet problems with your vet when in doubt – many encourage simple vinegar home remedies.

General pet tonic

Cider vinegar is recommended as a tonic for pets, given in small doses in the drinking water. The idea is to introduce the cider vinegar gradually so that the animal becomes accustomed to the slightly acidic flavour of the drinking water.
It is said to be beneficial to cats, dogs and rabbits as well as horses. Including cider vinegar in the diet deters ticks and fleas.

Many vets encourage an organic approach to pet care and will be happy to offer advice on how to supplement a particular pet's diet in this way.

1 Start by adding 1.5ml/¼tsp cider vinegar to the pet's drinking water.

2 Gradually increase the amount of vinegar until you are adding 5ml/1tsp.

Cider vinegar pet wash

Washing pets with cider vinegar is said to help prevent flea attacks. Check with your vet for expert advice on exactly the right proportion to use.

Add a few drops of vinegar to the pet's regular bath water. Make sure that you rinse it off well afterward with warm water.

Pet litter trays

Even after a thorough cleaning, plastic pet litter trays can sometimes retain unpleasant odours.

Rinse the pet litter tray out with white vinegar, scrub well with a brush and rinse again with the vinegar. Be sure to rinse the tray out well afterward in clean cold water – a fussy pet will refuse to use litter that still smells of vinegar. Leave to dry in the open air.

Horse grooming and care

Adding a little cider vinegar to foodstuff and/or grooming a horse with garlic vinegar is an effective way of keeping flies off. Vinegar can be used to clean the horse's legs after removing mud or to treat sweet itch, and can also help prevent the horse from scratching any irritating bites.

Pet hutches and cages

Rabbit hutches and the cages of other small pets, such as hamsters or mice, can be cleaned out with distilled white vinegar.

Clean the animal's hutch or cage as usual. Wipe out the inside of the hutch with white vinegar. This will neutralize any odours. Wipe the hutch down again, this time with clean water, to remove any lingering smell of vinegar. Allow to dry before replacing any bedding.

VINEGAR FOR THE CAR AND GARDEN

Cut down on costly, environmentally unfriendly chemicals by using vinegar outside your home, in the garden and to clean your car.

Flowerpots

Use vinegar to clean and remove build-ups of mould from flowerpots that have been left outside.

Scrub unglazed flowerpots with a little neat vinegar to remove any mould or growth. Wash thoroughly and allow to dry.

Greenhouse

Use vinegar to clean greenhouse glass.

Use a soft brush to brush vinegar on to the glass, then scrub with crumpled up newspaper. Wash down with water and rinse with a bucket of water to which 250ml/ 8fl oz/1 cup of vinegar is added.

Insect trap

A mixture of vinegar and apple juice will attract insects and keep them off plants.

Put 250ml/8fl oz/1 cup each of vinegar and apple juice in a container. Hang near plants you want to protect. Insects will be lured to the vinegar mixture and away from plants.

Insect deterrent

Use vinegar as an insect deterrent on potted plants where adding acid to the soil is not a problem.

Add a few drops of vinegar to the soil, taking care not to get any on the plant itself. Water the plant as usual.

Poles and plant supports

Prevent garden canes, poles and plant supports going mouldy while they are in storage.

Wipe down with neat vinegar when removing poles from plants. Leave to dry completely before storing.

Ant deterrent

Once you have rid a nest of ants, deter them from returning with vinegar.

Pour boiling water on an ants' nest and all around the surrounding area. Once dry, pour vinegar around the area to stop the ants returning.

Garden sprayers

Vinegar can be used to clean out weedkiller from the insides of a garden sprayer.

Add 250ml/8fl oz/1 cup each of water and vinegar to an empty garden sprayer after rinsing out the weedkiller. Shake thoroughly and spray the mixture through the nozzle to clear it. Leave to stand overnight, then rinse thoroughly. If in any doubt, read the instructions for the particular weedkiller used in the sprayer to check that it will be neutralized by the acid in the vinegar.

Weed deterrent

Use vinegar to clean paving slabs and as a weed deterrent around pathways and doorsteps.

Remove all weeds and brush the slabs down to remove all loose dirt. Add 475ml/16fl oz/2 cups vinegar to a bucket of warm water and use to scrub down paving slabs. If the slabs are especially stained, green or dirty, spray neat vinegar all over them and leave to soak before scrubbing down. Drizzle neat vinegar around the slabs and steps to prevent the weeds coming back.

Car windscreens

Use vinegar and toothpaste for a crystal-clear windscreen.

Wash the car windscreen down with warm water, then use a little toothpaste as a mild abrasive to clean the glass, then rinse off with water. Add 120ml/4fl oz/ ½ cup vinegar to a bucket of warm water, wring out a cloth and wipe the windscreen with the vinegar and water mixture. Polish with a chamois leather before it dries. Take care not to get vinegar on the car's paintwork as it may damage it.

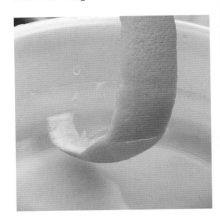

Cat repellent

To overcome a cat-spraying problem, use a combination of orange peel and vinegar.

Dip strips of orange peel in vinegar and place them around the cat's favourite spraying place. Keep spraying the peel with vinegar until the cat has moved on to a new place.

Slug barrier

A barrier of vinegar-soaked chippings around plants will deter slugs.

Soak wood chippings in vinegar and place around beds of seedlings. Maintain with vinegar top-ups. Do not flood the area – it will kill the plants.

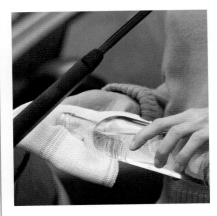

Windscreen wipers

Remove the greasy build up on windscreen wiper blades with vinegar.

Pour a little white vinegar on to a cloth. Clean the windscreen wipers with vinegar. This will remove grease and also stop them from squeaking.

VINEGAR IN THE KITCHEN

Various types of vinegar are used for all sorts of cooking. While many methods are steeped in history, others are comparative newcomers to the culinary scene. Whether it is used for marinating, dressing, deglazing, sharpening, flavouring, or even raising cakes, the addition of vinegar will enrich and improve many dishes. From essential food preservation techniques to frivolous finishing touches, vinegars of one sort or another should feature in every cook's repertoire.

Left: Preserve the essence of fresh fruits and herbs by making your own flavoured vinegars.

CULINARY USES OF VINEGAR

Versatile vinegar plays several essential roles in classic cuisines from around the world, from food cleaning and preparation, to flavouring, marinating and of course preserving.

Basic types of vinegar can be used throughout the processes of preparing, cooking, preserving and serving food. The array of fine-flavoured, and often expensive, vinegars now widely available are usually used to add the finishing touches to dishes, to flavour drinks or, quite simply, served as dipping sauces.

Vinegar in food preparation

Removing strong flavours Vinegar is used to rinse or soak food that naturally has an odour or for removing, or reducing, strong flavours that have developed when food has deteriorated slightly (but not to the extent of being inedible). For example, fresh skate has a slight smell of ammonia and washing in water acidulated with vinegar removes or reduces this. Rinsing with vinegar also counteracts slightly rancid flavours, for example if fat that is tainted is removed from poultry or meat, the meat can be rinsed in vinegar to remove any residual flavour. Venison fat (of which there is not usually very much) has a bad flavour and rinsing in vinegar after trimming off any fat and membrane removes any residual flavour. Vinegar is also used in the preparation of tripe to clean it and remove unpleasant flavours.

Cleaning Vinegar is used in the first stages of food preparation to clean produce. Adding a little vinegar to a bowl of water in which fresh garden lettuce and other leaves

are washed brings out insects and bugs that otherwise manage to conceal themselves throughout several rinses. The method also works well for wild berries, such as blackberries, that tend to harbour tiny maggots.

Vinegar is often used with salt to prepare snails. The freshly gathered snails are purged (fed on herbs or starved) for about a week in a box with a finely perforated lid to rid them of any remnants of poisonous plants. They are washed and then sprinkled with salt and vinegar. This process is repeated until the snails are free from slime, when they are boiled and shelled ready for using.

Preventing discolouration Placing peeled fruit and vegetables that tend to discolour in cold water with a little vinegar added prevents the items from

changing colour while the rest of the batch is prepared. For example, this is useful when preparing potatoes, artichokes, apples and pears.

Marinating and macerating Vinegar can be used to marinate savoury ingredients or macerate fruit. The choice of vinegar depends on the type of dish and whether it will be used in a sauce or discarded. The majority of marinades and soaking mixtures are either used in the cooking liquid, to baste food or served in a sauce.

When macerating fruit, the vinegar usually acts as a dressing or sauce, or it is used in a syrup. Fine wine vinegars or fermented fruit vinegars are ideal for macerating fruit to be served raw and the acidity of the vinegar brings out the flavour of fruit, such as strawberries or papaya, that do not have a high acid content.

Above: Adding vinegar to water when washing salad will remove insects.

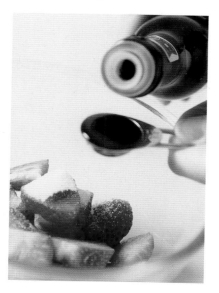

Above: The acidity in balsamic vinegar enhances the flavour of certain fruits.

Above: Vinegar is commonly used with oil as a dressing or finishing ingredient.

Above: Vinegar has been valued for its preservation properties for centuries.

DEGLAZING WITH VINEGAR

Vinegar in cooking

Balancing strong flavours A little vinegar is often added to the poaching liquid for strongly flavoured fish and is also used in court bouillon, a cooking liquid used for fish.

Sharpening One of the main contributions vinegar makes to food is to add a sharp flavour. When food is bland or without any hint of acidity, a little vinegar accentuates the flavour and brings out the characteristics of the ingredients, typically with fruit, such as blueberries, but also with vegetables, such as beetroot (beets). Flour-based sauces may be slightly sharpened with a hint of vinegar or it can be used to bring contrast to soups and gravies.

Cutting richness Vinegar is excellent for balancing the richness of dishes and foods that have a high fat content. Examples include egg-based sauces and in roasting juices or sauces for rich and fatty main foods, such as duck or lamb.

Setting eggs Adding vinegar to water when poaching eggs helps to set the whites quickly.

Raising agent The chemical reaction between vinegar and baking soda serves as a raising agent in some traditional baking recipes.

Vinegar in preservation

The preservation properties of vinegar are fundamental to its culinary value and it is for this purpose that vinegar is used in the largest proportion. It is used in a wide variety of preserves, with ingredients ranging from fish to fruit, and from simple pickles that require minimum preparation to more sophisticated jellies and store sauces.

Vinegar in finishing and dressing

Added to a cooking pan after frying or roasting food, vinegar is used to loosen the residue and incorporate the juices and some fat in a sauce or gravy (this is called deglazing). However, the simplest use of vinegar is as a straightforward dressing or condiment. It is drizzled over cooked or raw food or dishes to add piquancy and contrast. Partnered with oil it is used for a variety of dressings.

Cider or wine vinegars are often used to deglaze cooking pans, either after frying, grilling (broiling) or roasting.

Balsamic vinegar is probably one of the most popular vinegars for this process, but all sorts of vinegars are useful, including fruit vinegars, spiced and herb vinegars. The important thing is that the harshness of the vinegar balances the food or other ingredients that may be added to a sauce. Rich, mellow, sweet-sour vinegars can be used to deglaze the pan and complete the dish without the need for any other ingredients. On the other hand, deglazing with a sherry vinegar, brown rice vinegar or cider vinegar can bring just the right level of piquancy to a gravy or rich wine sauce.

When you are deglazing a pan, add comparatively small quantities of liquid in one go and keep scraping, whisking or stirring the sediment and juices off the pan.

Control the heat to avoid burning the sediment or the vinegar. This is particularly important when using rich and sweet vinegars, such as balsamic.

Sauces made by deglazing are often left concentrated in flavour and modest in volume, so a little is drizzled over each portion. Instead of coating food with a larger quantity of lighter sauce, this allows for pleasing contrasts when eating the main dish and vegetables or salad, for example, as flavours and textures come through individually.

VINEGAR IN DRESSINGS AND MARINADES

When used in dressings, vinegar adds a tangy kick to vegetables and salads. Meat, fish or poultry are tenderized by vinegar marinades, ready for cooking.

On its own or partnered with oils and herbs, vinegar is the classic dressing ingredient. There is nothing easier than drizzling rich balsamic vinegar over a green salad. Simple oil and vinegar salad dressings need little introduction – they are as familiar and popular as mayonnaise. Home-made vinegar dressings are easy to mix and keep well in the refrigerator.

Beyond the salad

Vinegars and dressings do not have to be reserved for salads – they are also terrific on grilled (broiled), pan-fried, stir-fried, steamed or poached foods. Use favourite vinegar dressings instead of rich milk or cream sauces or high-fat gravies to bring pleasing piquancy and contrast to fish, poultry or meat. Drizzle a little dressing over boiled new potatoes or cauliflower instead of butter or cheese sauce. Top with a few Parmesan shavings for a hint of cheese. Toss freshly boiled or steamed green or runner beans with a little dressing instead of loading them with butter.

Marinating with vinegar

Vinegar is used in marinades to tenderize meat. The vinegar may be used with water or wine and the proportion is usually relatively small. This type of marinade is popular for all meats and game (especially venison). The flavouring ingredients (spices, herbs, onion, carrot) are usually heated with a little water, wine or vinegar first so that they release their flavour, then allowed to cool. More marinating liquid may be added. When cooled, the food to be marinated is added and turned in, or basted with, the mixture.

Marinating time depends on the type of food, dish and intensity of flavour required – it can be anything from 30 minutes to 3 days. For short

VINEGAR DRESSING TIPS

When you are making a vinegar-based dressing, always start with the vinegar, dissolving all the flavourings and seasonings in the vinegar. Salt, sugar or mustard, for example, should be mixed into the vinegar first until dissolved. A good way to do this is to place all the ingredients together in a screw-top jar and then shake them up. Then the oil should be added. This way the seasonings sing through the whole dressing rather than separating out as sediment at the bottom.

Pick the vinegar and oil to suit the main ingredients in the dish, and to suit personal preferences – mild cider vinegar makes a super-soothing dressing compared to a more piquant wine vinegar.

Using vinegars flavoured with fruit or herbs can completely change the appearance of a salad, turning a simple plate of green leaves into a dish of flavour contrasts, for example.

For a well-balanced oil and vinegar mixture, you should use one part vinegar to two parts oil.

Not all dressings need oil. Try a simple dressing of rich balsamic vinegar or sweet raspberry vinegar, with a little citrus zest and some ground black pepper.

Above: Herb vinegars give extra flavour to salad dressings and marinades.

Above: Vinegars flavoured with garlic or spices make great salad dressings.

BASTING

Marinades should always be heated if they are used for basting. This avoids cross contamination from the raw, unheated mixture in which the raw meat has been marinating to the finished cooked food. It is particularly important when brushing a marinade over grilled (broiled) food, such as chicken or turkey.

If the marinade is not brought to the boil after marinating, bacteria from the raw food may be brushed over the grilled food just before serving.

Above: Marinating lamb chops in rich balsamic vinegar for several hours before they are cooked will leave the meat tender and moist.

MARINADE INGREDIENTS

Cider or wine vinegars are used for marinades: light ones for fish and poultry, red or darker ones for meat and game.

Fruit vinegars make excellent marinades, adding flavour interest. Some unpasteurized fruit vinegars (such as pineapple vinegar) have strong tenderizing properties if they retain the enzyme from the fruit that gradually breaks down protein.

Typical spices that are used for marinating red meat include juniper berries (good with game), coriander, cloves, mace (the blade of mace rather then the ground spice) and cinnamon sticks.

The most suitable herbs for marinating include rosemary, thyme, oregano and sage. Dill and fennel are often used in lighter marinades, with white wine or cider vinegar.

marinating the food is left covered in a cool place but not chilled; for longer marinating the food is placed in the refrigerator. When left for any length of time, the food should be turned occasionally or regularly to ensure that it is evenly flavoured.

The food may be drained before cooking, for example when grilling (broiling), roasting or barbecuing. The marinade may be used for basting or glazing during or after cooking, or it may be included in the cooking liquid, for example when making a casserole, braising or pot roasting. Alternatively, it may be added to, or heated as, a sauce at the end of cooking.

Matching vinegar with ingredients

Key points to consider are the strength of flavour of the food and any other ingredients, whether it is served raw or cooked and the required finished flavour, particularly in terms of tartness or sweet-sour balance.

Fitting the vinegar to the food means selecting according to how sharp it is, whether it has a clear 'brisk' sharpness without any supporting fruitiness, richness or sweetness.

While some flavour characteristics survive in hot dishes others are completely lost. The distinct sherry characteristics of some sherry vinegars are good in hot sauces and balsamic vinegar will clearly flavour roasted vegetables, but the flavour of blueberry vinegar or an aromatic moscatel vinegar is diminished drastically when used in cooking.

In macerating or marinating the pronounced flavours of some vinegars permeate and are balanced by main ingredients, but the delicate fruitiness of other vinegars is diluted. Fermented fruit vinegars (as opposed to fruit-flavoured vinegars) are perfect for drizzling over just before serving when their flavours will still be fresh and clear, without marinating or macerating, and no cooking.

When vinegars are drizzled over immediately before serving and not stirred in, even though they may be fruity or relatively delicate in flavour, they may well match foods or dishes that are quite heavily spiced or flavoured with herbs. Instead of being incorporated, they bring occasional contrast as the food is eaten.

VINEGAR AND PRESERVATION

One of the key culinary uses of vinegar has always been, and still is, as a preservative, especially for fruit and vegetables in pickles and chutneys.

One of the first uses of vinegar was as a food preservative. Being able to preserve food during times of plenty for periods of scarcity was essential for survival. Right up until modern technology provided practical, affordable alternatives by way of refrigerators and – especially – freezers, every cook relied on vinegar for preserving summer produce for winter meals. Preserves may no longer provide an essential source of sustenance for winter but they are an inherent part of the cuisines of all countries. They bring intense, contrasting flavours to simple meals, they enliven burgers, grills and barbecues and add a burst of flavour to sandwiches and filled breads.

How it works
Cooking foods with, or immersing them in, vinegar creates an acidic environment that is not suitable for the growth of the majority of micro-organisms that cause food spoilage. This is the easiest method of making preserves and it is used particularly for vegetables and fruit. The sharpness of the vinegar is often balanced by adding sugar, creating a sweet-sour flavour, and the high proportion of sugar in many preserves also helps to prevent spoilage.

Mature flavours
The majority of vinegar preserves should be allowed to mature before they are eaten. This allows time for the spices, herbs and other flavourings to penetrate

Above: The mild flavours of cider vinegar or white wine vinegar work well when pickling strong-tasting ingredients such as these aromatic Thai pink shallots.

the main ingredients. Individual flavours mingle and mellow, linked by the sharp or sweet pickle. Rich strong preserves that are intended to keep for up to 2–3 years taste better after 1–3 months' maturing, whereas others that taste best eaten within 6 months are usually ready after 2–3 weeks' maturing.

Keeping qualities
Ingredients which have been preserved in vinegar keep very well for long periods, with deterioration in texture or colour being the usual limiting factors rather than drying out or micro-organism attack. As long as

all the equipment used is thoroughly cleaned, the ingredients properly prepared and measured, and the cooking method followed, foods preserved in vinegar will keep for years. The jars or bottles have to be properly covered and stored in a cool dark place.

Before the food becomes mouldy or develops off flavours, it is likely to soften in texture (or, in the case of eggs become very rubbery) and look discoloured. The natural flavour of the ingredients may also diminish in intensity, so even though the food is safe to eat it will not taste its best.

Types of vinegar preserves

As a preservative, vinegar is used in savoury recipes, including pickles, chutneys, sauces, ketchups, relishes and jellies. Sugar is often used to counterbalance the sharp taste, giving a sweet-sour flavour.

Pickle This may be simple pickled ingredients, such as vegetables, preserved in plain or spiced vinegar. The basic ingredients may be raw or cooked. Additional flavouring ingredients can be added or the vinegar can be sweetened. When choosing vinegar for sweet pickles, it is important to think about the colour as well as the flavour of the vinegar. For green fruits such as green figs or red fruits such as plums, choose a light-coloured vinegar and spices or flavourings that won't alter the colour. Using a dark vinegar will turn the fruits a sludgy brown colour. Yellow fruits such as nectarines and apricots, and white or creamy fruits such as pears, look stunning pickled in a coloured vinegar such as raspberry. As well as vegetables, fish, eggs, fruit and unripe walnuts can be preserved by pickling.

Pickle is also the term used for a mixture of ingredients, usually cut into chunky pieces, and preserved in a vinegar-based sauce. The ingredients are usually cooked, either in the pickle liquid or before they are immersed in the sauce. Piccalilli is a classic example – it consists of mixed vegetables cooked in a sweet-sour mustard and turmeric sauce – but any chunky equivalent of a chutney could be called a pickle.

Chutney Chutney is the term for a cooked preserve. Onions, garlic, spices and other flavouring ingredients are combined with fruit and/or vegetables, then cooked in vinegar until thick and rich. Sugar is added to balance the flavour. As a rule, a chutney has a finer texture than a pickle and it is cooked for longer. Some chutneys are long cooked and well reduced.

Relish Relish refers to the sweet-sour and/or spicy flavour of the preserve, which is usually finer in texture than a pickle but not as smooth and thick as a chutney. Chilli and/or other hot spices may be used in relishes but some are mild, for example corn relish is sweet-sour and mild.

Jellies These are essentially fruit and sugar preserves, similar to jams and sweet jellies. Vinegar is used to make sweet-sour herb or fruit jellies to complement meat or cheese. Apple, lemon and mint jellies are typical.

Above: Brown malt vinegar is the best choice for use in rich chutneys.

Above: White vinegar can be bought ready-spiced or spiced at home for pickling.

Above: When pickling fruit, think about the colour as well as the taste of the vinegar.

CULINARY TRICKS WITH VINEGAR

Vinegar is useful in the kitchen for improving the performance of ingredients or enhancing simple mixtures in a wide variety of ways.

Poaching eggs

Adding vinegar to the poaching water makes the whites set more quickly.

INGREDIENTS

5ml/1 tsp vinegar

1 egg

English breakfast muffins, to serve

2 Poach the egg for 2–3 minutes in the barely simmering water. This will leave the yolk soft and runny. The circular swirl of water keeps the egg white in good shape, preventing it from spreading in strands, and the vinegar helps to set it quickly.

Peeling eggs

Adding a little vinegar to water makes it easier to peel hard-boiled eggs when they are cooked.

Add 5ml/1 tsp vinegar and 15ml/1 tbsp salt to boiling water before gently placing the eggs into the water.

1 Add 5ml/1 tsp vinegar to a pan of water and bring to the boil. Reduce to a simmer. Crack an egg into a cup, swirl the water with a spoon, then slip the egg into the middle of the swirl.

3 Carefully remove the egg from the water using a slotted spoon.

4 Serve immediately with buttered English breakfast muffins.

Making eggs go further

This is an old-fashioned suggestion for making one egg do the work of two.

When baking a cake, use 15ml/1 tbsp vinegar with 1 egg instead of adding 2 eggs.

Making salad cream
From a 1930s cookbook by Elizabeth Craig, this was suggested as a good alternative to mayonnaise.

Pour a small can of evaporated milk into a large bowl and alternately mix in 250ml/8fl oz/1 cup vinegar and 30ml/2 tbsp vegetable oil. Add a little salt, pepper and mustard to taste.

Using vinegar in sweet dishes
Use sweetened fruit vinegars and balsamic vinegar to dress fruit for dessert. Serve alone or with cream or yogurt.

Drizzle balsamic vinegar over pan-fried or grilled (broiled) fruit, such as bananas, peaches or pears. Serve fruit vinegars with plain vanilla mousse, cheesecake or ice cream for a tangy contrast.

Storing cheese
Wrapping cheese in a vinegar-soaked cloth before storing it in the refrigerator will keep it fresh and moist.

Place a little vinegar in a bowl. Soak a clean cloth or paper towel in the vinegar. Squeeze out most of the liquid, then use to wrap the cheese. Place in an airtight container and store in the refrigerator.

Preserving colour of vegetables
Adding a little vinegar to the water will improve the colour of boiled vegetables.

Bring a large pan of water to the boil, then add 15ml/1tbsp vinegar before carefully placing the vegetables in the water and cooking as usual. This will help them to retain their colour.

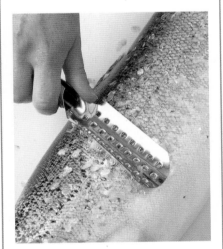

Descaling fish
Rubbing fish skin with vinegar will make it easier to remove the scales.

Place a little vinegar in a small bowl and rub it into the fish skin with your fingers. Leave for 5 minutes, then descale the fish as usual. The vinegar will also help to reduce any fishy odour.

Cooking pasta
Adding vinegar to the cooking water will prevent pasta from being sticky.

Bring a large pan of water to the boil. Instead of adding salt, add 5ml/1tsp vinegar. Add the pasta and cook as usual. The addition of the vinegar will reduce the starch and make the pasta less sticky and easier to serve.

FLAVOURING VINEGARS

A wide variety of commercially produced flavoured vinegars are available, but it is easy and fun to try making your own. Use vegetables, herbs, spices and even flowers.

FLAVOURING VINEGARS AT HOME

There are endless possibilities for flavouring vinegars at home, from spicing inexpensive vinegar for pickling to flower-scented vinegars and vibrant fruit vinegars. Flavouring vinegar is not that difficult to do at home and it is a good way of preserving the essence of fresh garden herbs or local seasonal fruit.

Experiment with fresh garden produce, lively herbs and favourite spices to make a range of vinegars that can be used to infuse exciting flavours into salad dressings and sauces.

The choice of vinegar depends on its likely uses. Malt may be ideal for pickles and chutneys but not practical for salad dressings, sauces or marinades. Wine vinegar will go well in some dressings or for deglazing a cooking pan but may be too harsh for dressing a fruit dessert.

Cider vinegar will be useful for dressings, sauces and drinks. The recipes here are a mixture of traditional and modern, and intended to spark off ideas for experimenting with flavouring and using a variety of different vinegars.

Chilli vinegar

Easy to make with dried or fresh chillies, this chilli vinegar can be added to many dishes to give them that extra bite. It makes a great addition to pickles, chutneys and sauces, and will add a touch of fiery colour when drizzled on to salads, pasta dishes and pizzas.

INGREDIENTS

1–2 dried red chillies, or 1–3 fresh green or
   red chillies, halved and deseeded
1 bottle of vinegar

1 Pour the vinegar out of the bottle and reserve. Place the dried or fresh chillies in the bottle then refill with the vinegar. Seal.

2 Leave the chilli vinegar to infuse in a cool, dark place for anything from a week to 6 months.

Variation

Deseed and finely dice 1 red (bell) pepper and add it to the vinegar with the chillies to make a fiery pepper chilli vinegar.

Cucumber vinegar

The fresh taste of cucumber makes this a perfect vinegar to use in salad dressings.

Dice a cucumber and layer it in a bowl with a light sprinkling of salt. Leave for at least 3 hours to draw out the excess water. Drain and squeeze out the cucumber, then place in a clean dry bowl. Pour in enough vinegar to cover generously. Cover and leave to stand for 3 days before straining and bottling.

Shallot vinegar

A traditional accompaniment to oysters, this is best made with wine vinegar.

Add a finely diced shallot to a bottle of vinegar. Seal and infuse (steep) for 1–2 weeks, then strain and bottle. Alternatively, flavour bottles of vinegar with trimmed and chopped spring onions (scallions) or chives. Chinese garlic chives can also be used to make a strongly flavoured vinegar.

Ginger vinegar

Use this in oriental cooking, in stir-fried dishes or dipping sauces.

Finely chop a piece of fresh root ginger, about 4cm/1½in. Simmer the ginger in 5cm/2in vinegar in a covered pan for 30 minutes. Cool before transferring to a glass container, then pour on enough vinegar to fill. Seal and infuse for about a month, then strain and add sugar to taste before bottling.

Horseradish vinegar

Add a little horseradish vinegar to gravy and serve with roast beef.

Scrub and grate 115g/4oz fresh horseradish. (Horseradish is very strong so wear gloves.) Put the grated horseradish in a glass bottle and cover with vinegar. Seal and infuse (steep) for at least a week before straining and bottling.

Celery vinegar

This can be added to soups, sauces or salad dressings.

Slice 1–2 celery stalks lengthways. Place the celery in a jar, without packing in tightly. Pour in vinegar to cover, shaking out all the air, cover and leave to stand for 3–4 weeks. Strain and bottle the vinegar.

Garlic vinegar

Use this to make delicious marinades for fish, shellfish and chicken.

Peel and halve one large or two small cloves of garlic, then add the garlic to a bottle of vinegar. Seal, and leave to stand for at least a week, then strain and bottle the vinegar.

Herb vinegars

A wide variety of herbs can be used to flavour vinegar, either individually or mixed. Tarragon is the traditional choice – tarragon vinegar is a classic ingredient for salad dressings and egg sauces.

INGREDIENTS
about 15g/1/2oz/³/₄ cup fresh tarragon
450ml/³/₄ pint/scant 2 cups vinegar

1 Wash and dry the tarragon and remove the leaves from the stalks.

2 Place in a container and pour on the vinegar. Seal and leave to stand for 4 weeks in a cool, dark place. Strain, bottle and cover.

HERB VINEGARS

• Mint, basil, bay leaves, dill, thyme, marjoram and rosemary all impart a full flavour to wine or cider vinegars for use in sauces and dressings as well as for pickling.

• Cider vinegar is best for making herb vinegar. White or red wine vinegars are also acceptable, but malt vinegar, including light or white, is too harsh and strongly flavoured for herbal use.

• Herbs used for making flavoured vinegars should be picked in the morning after the dew has dried, but before the flavours have dissipated in the hot sun. Allow the moisture to dry completely before use, otherwise they may become mouldy.

• Herb vinegars should be strained, but you can add a few fresh sprigs of herb when bottling to indicate the flavour.

• Attractive bottles of herb vinegars make excellent gifts. Look out for interesting bottles to decant your vinegar into and use ribbons and labels for decoration.

• Try the following combinations to flavour white or red wine vinegar or cider vinegar:
    Sage, thyme, bay and marjoram
    Sage, thyme and basil
    Sage, dill and coriander (cilantro)
    Basil, thyme and oregano
    Lemon grass and rosemary
    Coriander, chives, dried
      chillies and garlic

Rose petal vinegar

For a really decadent dressing, make this with Champagne vinegar.

Pull the rose petals from 2 or 3 scented rose heads. Snip off the bitter white heel at the base of each petal. Place 2 handfuls of petals in a large glass jar or bottle. Fill the jar or bottle with Champagne or white wine vinegar, seal very tightly with a screw-top or cork and leave on a sunny window sill for at least 3 weeks before straining.

Nasturtium flower vinegar

Serve this as a dressing for salads containing nasturtium flowers.

Place a handful of nasturtium flowers in a jar, filling it without squashing the flowers. Pour in wine vinegar to cover, gently agitate the jar to remove trapped air and cover. Leave to mature for 2 months. Strain the vinegar through muslin (cheesecloth), squeezing out all the vinegar from the flowers.

Pickled nasturtium seeds

The seeds can be used as a flavouring ingredient in a similar way to capers.

Layer 75g/3oz nasturtium seeds in a large bowl and sprinkle with salt. Leave for 24 hours, then rinse and drain well. Dry the seeds well on a dish towel and place them in clean jars without packing them in too tightly. Pour in vinegar to cover. Seal and allow the seeds to mature for at least 2 months before using both the vinegar and the seeds.

MAKING FLOWER VINEGARS

You should always be certain that flowers are edible before putting them to food use in vinegars as some flowers can be poisonous. Common edible flowers include nasturtium, rose, pansy, citrus blossom, lilac, violet and honeysuckle.

Whenever using flowers for flavouring or cooking, it is best to pick them in dry weather and early in the day, when the flowers have opened but have not had time to wilt in the sunshine. Select perfect examples and shake off any insects. Try to avoid washing the flowers as this can bruise and damage them.

Make these flower vinegars with mild flavoured vinegars – use white wine vinegar or cider vinegar for the best results.

Pickled nasturtium buds

These buds have a peppery taste that is great in salads.

Pick 15–20 buds in dry weather and lay them on a paper towel for 3 days in a cool dry place. Place in jars and cover with boiling vinegar. Allow to mature for 3 months before using. Both buds and vinegar can be used.

Lavender vinegar

This is delicious when used in dressings or drizzled over roast meats.

Trim the ends off several stalks of lavender and add to a bottle of vinegar. Shake to dislodge air bubbles, cover and leave to stand for 6 weeks. Strain through muslin (cheesecloth) and store in clean bottles.

Raspberry vinegar

One of the favourites, this is simple to make with fresh or frozen raspberries. There is little point in buying raspberry flavoured vinegar when it can be made so easily at home using a favourite wine or cider vinegar and just the right hint of sweetness.

INGREDIENTS

450g/1lb raspberries, fresh or frozen

1.2 litres/2 pints/5 cups vinegar

1 Macerate the raspberries in the vinegar for about 5 days.

2 Strain through a sieve (strainer) lined with muslin (cheesecloth).

3 Heat with 450g/1lb sugar until dissolved, then bring to the boil. Bottle, cover immediately and cool.

FRUIT-FLAVOURED VINEGARS

These can be made in modest amounts using fresh, frozen or canned fruit. Leave unsweetened or add sugar to taste. When there is a glut of fresh fruit to use, then large quantities can be prepared, allowing about 2.25kg/5lb fruit to 2 litres/3½ pints vinegar.

White vinegar can be used for preparing large quantities but cider and wine vinegars can give better flavours and are practical for making smaller quantities. The fruit should be left to stand in the vinegar for 5–7 days before straining. When making small quantities to be used relatively quickly, there is no need to heat the vinegar. When preparing a large volume for long-term storage, the vinegar should be boiled before storing otherwise it will ferment with time.

Brandied raspberry vinegar

The following old-fashioned recipe makes a rich raspberry vinegar.

Measure the raspberries by volume and add twice the volume of vinegar to the raspberries. Leave to stand for a day, then strain. Add the same quantity of raspberries and leave to stand for a further day. Strain again into a large pan and add 450g/1lb sugar to every 600ml/1 pint/2½ cups raspberry vinegar. Bring to the boil, stirring until the sugar has dissolved, reduce the heat and simmer for 1 hour. Skim off any scum as the vinegar simmers. Measure the reduced vinegar and add about 250ml/8fl oz/1 cup brandy to every 600ml/1 pint/2½ cups vinegar.

Rich blackberry vinegar

Mixed with sparkling water and a sprig of mint, this makes a refreshing drink.

Measure a jug (pitcher) of blackberries and add the same volume of vinegar. Cover and stand for 1 day. Strain, add another jug of berries, cover and leave for a further day. Strain and repeat again. Strain and add 450g/1lb sugar for every 600ml/1 pint/2½ cups vinegar. Heat until dissolved, then bring to the boil. Reduce the heat and simmer for 30 minutes. Bottle and allow to mature for 4 weeks before using.

Simple blackberry vinegar

Use this method if you don't want to wait for the vinegar to infuse.

Place 100g/3¾oz fresh blackberries in a bowl, add 600ml/1 pint/2¾ cups vinegar, then leave for 2 days. Strain and transfer to a pan. Add 450g/1lb sugar to the vinegar, heat until dissolved, then bring to the boil. Simmer until the vinegar is reduced by half. Bottle.

Stone (pit)-fruit vinegar

This can be made with any stone-fruit, such as sloes, damsons or plums.

Use 100g/3¾oz. To break down the skins and flesh slightly, place the fruit in the freezer overnight. Pit the fruit and chop the flesh then place in a pan, pour on 600ml/1 pint/2½ cups vinegar and bring to the boil. Cool and leave to stand for 1 week, crushing the fruit with a vegetable masher. Strain and add a little sugar to sweeten if required before bottling.

Strawberry vinegar

This has a rich and fruity flavour and is great drizzled over ice cream.

Follow the same method as for brandied raspberry vinegar (see previous page), substituting strawberries for the raspberries and without adding the brandy at the end. Adding several batches of fruit gives the vinegar an excellent rich flavour.

Citrus vinegar

This tangy vinegar makes a great marinade ingredient for fish dishes.

Pare off the thin outer rind of an orange, lemon or lime using a vegetable peeler. All the white pith should be removed as it will taste bitter. Place a few strips of the peel in a bottle of vinegar and allow to infuse for 3–6 weeks. Using a mixture of different coloured fruits will give an attractive result. Alternatively, add the grated zest of 1 lemon, lime or orange to a bottle of vinegar.

EASY FRUIT VINEGARS

Prepare the fruit, discarding inedible parts, and cut into small pieces. Try peaches, nectarines, apricots, cherries or mango. Place in a glass mixing bowl and pour in vinegar to cover. Cover with cling film (plastic wrap) and macerate in the refrigerator for up to a week, stirring daily. Strain and sweeten the vinegar. Instead of discarding the fruit, cook down to a chutney or pickle with onions, spices, vinegar and sugar.

VINEGAR IN DRESSINGS

Vinegar is an essential ingredient in all sorts of dressings, whether you wish to add an edge to creamy dressings or a sharpness to classic combinations of herbs and oils.

Classic vinaigrette

Make sure all the ingredients are at room temperature to make a smooth emulsion.

Put 30ml/2 tbsp balsamic vinegar in a bowl with 10ml/2 tsp Dijon mustard, salt and ground black pepper. Whisk to combine. Slowly drizzle in 90ml/6 tbsp olive oil, whisking until the vinaigrette is smooth and blended.

Garlic raspberry dressing

Adding a splash of raspberry vinegar to this dressing enlivens a simple salad.

Fry 2 finely sliced garlic cloves in 45ml/ 3 tbsp olive oil until just golden. Remove with a slotted spoon and drain on kitchen paper. Pour the oil into a bowl and whisk in the raspberry vinegar. Season with salt and ground black pepper.

Hazelnut dressing

Serve this simple dressing with a goat's cheese salad and chopped hazelnuts.

Put 30ml/2 tbsp hazelnut oil in a small bowl. Add 5–10ml/1–2 tsp sherry vinegar or good wine vinegar to taste, and whisk the oil and vinegar together thoroughly. Season to taste with salt and ground black pepper.

Creamy raspberry dressing

Raspberry vinegar gives this quick dressing a refreshing, tangy fruit flavour.

Mix 30ml/2tbsp raspberry vinegar with 2.5ml/ 1/2tsp salt and stir until dissolved. Stir in 5ml/ 1 tsp Dijon-style mustard and 60ml/4 tbsp natural (plain) yogurt.

Dill dressing

Serve this simple mixture of oil, vinegar and dill with smoked fish.

Whisk 90ml/6 tbsp olive oil and 30ml/ 2 tbsp white wine vinegar together, or shake in a screw-top jar. Blend in 15ml/ 1 tbsp chopped fresh dill and season.

Creamy orange dressing

This tangy orange dressing is delicious with a mixed green salad.

Pour 45ml/3 tbsp crème fraîche and 15ml/ 1 tbsp white wine vinegar into a screw-top jar with the grated rind and juice of 1 orange. Shake until combined, then season to taste.

VINEGAR IN MARINADES

Using vinegar in marinades helps to tenderize meat and fish and bring out the flavours of other ingredients. Use a simple herb-flavoured vinegar or try these recipes.

Summer herb marinade

Raid the herb garden to make this fresh-tasting marinade.

Mix 45ml/3 tbsp tarragon vinegar, 1 crushed garlic clove and 2 finely chopped spring onions (scallions). Whisk in 90ml/ 6 tbsp olive oil, then add a large handful of chopped fresh herbs such as thyme or parsley and mix well. Season.

Hoisin marinade

Use this Chinese marinade for chops and chicken pieces.

In a small bowl, combine 175ml/6fl oz/ 1 cup hoisin sauce, 30ml/2 tbsp each sesame oil, dry sherry and rice vinegar, 4 chopped garlic cloves, 2.5ml/½ tsp soft light brown sugar and 1.5ml/¼ tsp five-spice powder.

Red chilli marinade

Spice up meat and fish with this hot and fiery marinade.

Mix 2 chopped, deseeded fresh red chillies with 1 clove crushed garlic, 1 tsp ground black pepper, 60ml/2fl oz white wine vinegar, 15ml/1 tbsp caster (superfine) sugar, 30ml/2 tbsp sunflower oil and 15ml/1 tbsp freshly chopped parsley.

Teriyaki marinade

In this traditional Japanese recipe, soy sauce is paired with rice vinegar.

In a small bowl, combine 1 crushed garlic clove, 5ml/1 tsp grated fresh root ginger, 30ml/2 tbsp dark soy sauce and 15ml/ 1 tbsp rice vinegar.

Thyme marinade

The cider vinegar in this herby marinade brings out the full flavour of the thyme.

Combine 90ml/6 tbsp olive oil with 30ml/ 2 tbsp cider vinegar, 1 crushed garlic clove, 5ml/1 tsp dried thyme and 2.5ml/½ tsp crushed peppercorns.

Sherry vinegar marinade

This piquant mix of sherry vinegar and garlic oil would suit fish or meat.

Mix together 30ml/2 tbsp sherry vinegar with 45ml/3 tbsp garlic-infused oil and season with salt and ground black pepper to taste.

VINEGAR IN SAUCES

Vinegar serves a sharpening and/or mixing function to add piquancy or bring flavours together effectively in many classic sauces.

Tartare sauce

This creamy sauce is the traditional accompaniment to fish dishes.

INGREDIENTS

Makes about 475ml/16fl oz/2 cups

1 egg yolk

15ml/1 tbsp white wine vinegar

30ml/2 tbsp Dijon-style mustard

250ml/8 fl oz/1 cup groundnut (peanut) oil

30ml/2 tbsp fresh lemon juice

45ml/3 tbsp finely chopped spring
  onions (scallions)

30ml/2 tbsp chopped drained capers

45ml/3 tbsp finely chopped sour dill pickles

45ml/3 tbsp finely chopped fresh parsley

1 In a bowl, beat the egg yolk with a wire whisk. Add the white wine vinegar and mustard and continue to whisk for about 10 seconds. Whisk the oil into the mixture in a slow, steady stream.

2 Add the lemon juice, spring onions, capers, sour dill pickles and parsley and mix well. Keep chilled; use within 2 days.

Energy 1662kcal/6835kJ; **Protein** 5.3g; **Carbohydrate** 2.7g, of which sugars 2.4g; **Fat** 181.2g, of which saturates 19.8g; **Cholesterol** 202mg; **Calcium** 141mg; **Fibre** 3.2g; **Sodium** 29mg.

Mint sauce

This sauce, with its tangy vinegar taste, is usually served with roast lamb.

INGREDIENTS

Makes about 300ml/^{1}/2 pint/1^1/4 cups

1 large bunch mint, finely chopped

105ml/7 tbsp boiling water

150ml/^{1}/4 pint/^{2}/3 cup wine vinegar

30ml/2 tbsp sugar

1 Place the chopped mint in a 600ml/1-pint/2^1/4-cup jug (pitcher).

2 Pour the boiling water slowly into the jug, covering the mint, and then leave it to infuse for about 10 minutes.

3 When the mint infusion has cooled until it has reached a lukewarm temperature, stir in the wine vinegar and sugar.

4 Continue stirring (but do not mash up). Mint sauce will keep for up to 6 months stored in the refrigerator, but it is best used within 3 weeks.

Energy 161kcal/685kJ; **Protein** 3.9g; **Carbohydrate** 36.6g, of which sugars 31.3g; **Fat** 0.7g, of which saturates 0g; **Cholesterol** 0mg; **Calcium** 226mg; **Fibre** 0g; **Sodium** 17mg.

Horseradish sauce

This has a very strong flavour that complements steak or roast beef.

INGREDIENTS

Makes about 200ml/7fl oz/scant 1 cup

45ml/3 tbsp horseradish root

15ml/1 tbsp white wine vinegar

5ml/1 tsp sugar

pinch of salt

150ml/^{1}/4 pint/^{2}/3 cup thick double
  (heavy) cream, for serving

1 Peel the horseradish root and grate it finely in a food processor. Horseradish is powerful, so submerge it in water while you peel it. Place the grated horseradish in a bowl, then wash your hands.

2 Add the white wine vinegar to the bowl, with the sugar and the salt. Stir until thoroughly combined.

3 Pour into a sterilized jar. It will keep in the refrigerator for 6 months. A few hours before serving the sauce, stir in the cream.

Energy 774kcal/3190kJ; **Protein** 2.8g; **Carbohydrate** 9.9g, of which sugars 9.8g; **Fat** 80.7g, of which saturates 50.1g; **Cholesterol** 206mg; **Calcium** 98mg; **Fibre** 1.1g; **Sodium** 40mg.

Beurre blanc

A reduction of white wine and vinegar forms the base for this rich sauce.

INGREDIENTS

Makes about 150ml/^{1}/4 pint/^{2}/3 cup

3 shallots, very finely chopped

45ml/3 tbsp dry white wine or court-bouillon

45ml/3 tbsp white wine vinegar or
   tarragon vinegar

115g/4oz/^{1}/2 cup chilled unsalted
   butter, diced

lemon juice (optional)

salt and ground white pepper

1 Put the shallots in a pan with the wine or court-bouillon and vinegar. Gently bring to the boil and cook over a high heat until only about 30ml/2 tbsp liquid remains in the pan.

2 Remove the pan from the heat and leave to cool until the reduced liquid is lukewarm.

3 Whisk in the butter, one piece at a time, to make a pale, creamy sauce. Taste, then season and add a little lemon juice to taste, if you like.

4 If you are not serving the sauce immediately, keep it warm in the top of a double boiler set over simmering water.

Energy 869kcal/3576kJ; **Protein** 1.3g; **Carbohydrate** 4.7g, of which sugars 3.4g; **Fat** 94.1g, of which saturates 62.1g; **Cholesterol** 265mg; **Calcium** 32mg; **Fibre** 0.8g; **Sodium** 39mg.

Hollandaise sauce

This is an essential part of Eggs Benedict, but is also delicious served on vegetables.

INGREDIENTS

Makes about 135ml/4fl oz/^{1}/2 cup

115g/4oz/^{1}/2 cup unsalted butter

2 egg yolks

15–30ml/1–2 tbsp white wine vinegar or
   tarragon vinegar

salt and ground white pepper

1 Melt the butter in a small pan over a medium heat. Put the egg yolks and vinegar in a small bowl. Add salt and pepper and whisk until the mixture is completely smooth.

2 Slowly pour the melted butter in a steady stream on to the egg yolk mixture, beating vigorously the whole time with a wooden spoon to make a smooth, creamy sauce.

3 Taste the sauce and add more vinegar, and season, if necessary. Serve.

Cook's tip

Instead of adding the butter to the egg mixture in the bowl, you could put the mixture in a food processor and add the butter through the feeder tube, with the motor running.

Energy 874kcal/3596kJ; **Protein** 5.8g; **Carbohydrate** 0.6g, of which sugars 0.6g; **Fat** 94.3g, of which saturates 56.4g; **Cholesterol** 580mg; **Calcium** 60mg; **Fibre** 0g; **Sodium** 27mg.

Béarnaise sauce

This rich buttery sauce is often served with 'steak frites' in France.

INGREDIENTS

Makes about 300ml/^{1}/2 pint/1^1/4 cups

90ml/6 tbsp white wine vinegar

12 black peppercorns

2 bay leaves

2 shallots, finely chopped

4 fresh tarragon sprigs

4 egg yolks

225g/8oz/1 cup unsalted butter, diced and
   warmed to room temperature

30ml/2 tbsp chopped fresh tarragon

salt and freshly ground white pepper

1 Put the white wine vinegar, peppercorns, bay leaves, shallots and tarragon sprigs in a pan and simmer until reduced to 30ml/2 tbsp. Strain through a fine sieve (strainer).

2 Beat the egg yolks with salt and ground white pepper in a heatproof bowl. Stand over a pan of gently simmering water, then beat the strained vinegar into the yolks.

3 Beat in the butter, one piece at a time. Beat the tarragon into the sauce and remove from the heat. It should be smooth, thick and glossy.

Energy 1919kcal/7900kJ; **Protein** 14.2g; **Carbohydrate** 1.4g, of which sugars 1.1g; **Fat** 206.4g, of which saturates 127.8g; **Cholesterol** 1324mg; **Calcium** 227mg; **Fibre** 2.5g; **Sodium** 1740mg.

Sweet chilli sauce

This Thai sauce is great served as a dipping sauce with spring rolls.

INGREDIENTS

Makes about 350ml/12fl oz/1¹/₂ cups

6 large red chillies

60ml/4 tbsp rice vinegar or wine vinegar

250g/9oz caster (superfine) sugar

5ml/1 tsp salt

4 garlic cloves, chopped

10ml/2 tsp fish sauce

1 Cut the red chillies in half lengthways, then carefully remove the seeds with a sharp knife. Soak the seeds in hot water for about 15 minutes.

2 Chop the chilli flesh and place in a food processor. Drain the seeds and add them to the food processor. Add the vinegar, sugar, salt and garlic.

3 Blend in the food processor until the mixture is smooth, then transfer to a small pan and cook over a medium heat for about 20 minutes, or until the sauce has thickened.

4 Leave the sauce to cool, then stir in the fish sauce. The sweet chilli sauce can be stored in the refrigerator in an airtight container for 1–2 weeks.

Energy 1032kcal/4398kJ; **Protein** 6.7g; **Carbohydrate** 265.8g, of which sugars 262.4g; **Fat** 0.8g, of which saturates 0g; **Cholesterol** 0mg; **Calcium** 174mg; **Fibre** 0.8g; **Sodium** 2275mg.

Chilli sauce

Serve this fiery sauce as a dip with raw vegetables or tortilla chips.

INGREDIENTS

Makes about 250ml/8fl oz/1 cup

15ml/1 tbsp olive oil

1 small onion, finely chopped

1 garlic clove, crushed

200g/7oz can chopped tomatoes

1 fresh red chilli, seeded and finely chopped

15ml/1 tbsp balsamic vinegar

15ml/1 tbsp chopped fresh coriander (cilantro), optional

salt and ground black pepper

1 Cook the onion and garlic in the oil for 5–10 minutes, or until soft.

2 Add the chopped tomatoes to the pan, with their juice. Stir in the chopped chilli and balsamic vinegar.

3 Cook gently for 10 minutes, or until the mixture has reduced and thickened.

4 Check the seasoning, and adjust if necessary. Stir in the chopped coriander, if using, and serve hot.

Cook's tip

If you like your chilli sauce really fiery, use two red chillies instead of one.

Energy 165kcal/684kJ; **Protein** 3.3g; **Carbohydrate** 11.5g, of which sugars 10.1g; **Fat** 12.1g, of which saturates 1.7g; **Cholesterol** 0mg; **Calcium** 67mg; **Fibre** 3.6g; **Sodium** 27mg.

Coconut vinegar sauce

The main ingredient of coconut vinegar gives this sauce a real Filipino taste.

INGREDIENTS

Makes about 150ml/¹/₄ pint/²/₃ cup

60–75ml/4–5 tbsp coconut vinegar

3 red chillies, seeded and finely chopped

4 spring onions (scallions), finely chopped

4 garlic cloves, finely chopped

Place all the ingredients together in a bowl and mix thoroughly. Spoon into a jar, cover and store in the refrigerator for up to 1 week.

Energy 41kcal/171kJ; **Protein** 4.1g; **Carbohydrate** 4.9g, of which sugars 1.9g; **Fat** 0.7g, of which saturates 0.1g; **Cholesterol** 0mg; **Calcium** 37mg; **Fibre** 1.4g; **Sodium** 8mg.

Redcurrant Sauce

Serve this quick and easy redcurrant sauce with roast meats.

INGREDIENTS

Makes about 150ml/¹/₄ pint/²/₃ cup

115g/4oz/1 cup fresh or frozen redcurrants

10ml/2 tsp clear honey

5ml/1 tsp balsamic vinegar

30ml/2 tbsp finely chopped mint

Place all the ingredients in a bowl and mash them together with a fork. Store in the refrigerator and use within 2 days.

Energy 74kcal/316kJ; **Protein** 2.2g; **Carbohydrate** 16.8g, of which sugars 15.2g; **Fat** 0.2g, of which saturates 0g; **Cholesterol** 0mg; **Calcium** 133mg; **Fibre** 4.1g; **Sodium** 9mg.

Mustard and dill sauce

This sauce is traditionally served with salmon gravadlax.

INGREDIENTS

Makes about 200ml/7fl oz/scant 1 cup
1 egg yolk
30ml/2 tbsp brown French mustard
2.5–5ml/^{1}/$_2$–1 tsp soft dark brown sugar
15ml/1 tbsp white wine vinegar
90ml/6 tbsp sunflower or vegetable oil
30ml/2 tbsp finely chopped fresh dill
salt and ground black pepper

1 Put the egg yolk in a small bowl and add the mustard with a little soft brown sugar to taste. Beat with a wooden spoon until smooth.

2 Stir in the white wine vinegar, then whisk in the sunflower or vegetable oil, drop by drop to begin with.

3 As the sauce gets thicker, the oil can be poured in a steady stream. As the oil is added, the dressing will start to thicken and emulsify.

4 When all of the oil has been completely amalgamated into the sauce, season.

5 Stir in the chopped fresh dill. Cover and chill for 2 hours before serving.

Energy 716kcal/2950kJ; **Protein** 5.9g; **Carbohydrate** 6.3g, of which sugars 5.6g; **Fat** 74.3g, of which saturates 8.6g; **Cholesterol** 202mg; **Calcium** 106mg; **Fibre** 1.5g; **Sodium** 89mg.

Barbecue sauce

This sauce will enliven burgers and other food cooked on the barbecue.

INGREDIENTS

Makes about 500ml/17fl oz/generous 2 cups
30ml/2 tbsp vegetable oil
1 large onion, chopped
2 garlic cloves, crushed
400g/14oz can tomatoes
30ml/2 tbsp Worcestershire sauce
15ml/1 tbsp white wine vinegar
45ml/3 tbsp honey
5ml/1 tsp mustard powder
2.5ml/^{1}/$_2$ tsp chilli seasoning or mild
  chilli powder
salt and ground black pepper

1 In a large pan, heat the oil and fry the onion and garlic until soft. Stir in the remaining ingredients and gently bring to the boil.

2 Simmer the sauce, uncovered, for 15–20 minutes, stirring occasionally. Cool slightly. Pour the sauce into a food processor or blender and process until smooth.

3 Leave the sauce as it is, or press through a sieve (strainer) for a smoother result. Adjust the seasoning before serving. Store in the refrigerator and use within 2 weeks.

Energy 572kcal/2397kJ; **Protein** 9.1g; **Carbohydrate** 83.7g, of which sugars 73.6g; **Fat** 25.4g, of which saturates 2.7g; **Cholesterol** 0mg; **Calcium** 197mg; **Fibre** 9.6g; **Sodium** 413mg.

Lime butter sauce

Serve this creamy sauce as a dip for vegetables such as asparagus.

INGREDIENTS

Makes about 400ml/14fl oz/1^2/$_3$ cups
90ml/6tbsp dry white wine
90ml/6 tbsp white wine vinegar
3 shallots, finely chopped
225g/8oz/1 cup very cold unsalted
  butter, diced
juice of 1 lime
salt and ground black pepper
lime wedges, to serve

1 Mix the wine and the vinegar in a stainless steel pan with the chopped shallots, and simmer until the liquid has reduced down to about 15ml/1 tbsp.

2 Remove the pan from the heat, then vigorously whisk in the butter until the sauce thickens.

3 Whisk in the lime juice until it is thoroughly combined. Taste the sauce and adjust the seasoning if necessary.

Cook's tip

Do not allow the sauce to boil. If the butter is taking a long time to melt, put the pan back over a gentle heat to speed up the process.

Energy 1739kcal/7156kJ; **Protein** 1.9g; **Carbohydrate** 5.3g, of which sugars 3.9g; **Fat** 183.9g, of which saturates 121.5g; **Cholesterol** 518mg; **Calcium** 57mg; **Fibre** 0.8g; **Sodium** 88mg.

VINEGAR IN SOUPS

Vinegar brings a tangy contrast to many classic full-flavoured soups. The addition of a little wine vinegar or herb vinegar also brightens up rich, mild-flavoured soups.

Garlic soup

A little wine vinegar adds contrast to the rich and creamy garlic flavour.

INGREDIENTS

Serves 8

12 large garlic cloves, peeled and crushed

15ml/1 tbsp olive oil

15ml/1 tbsp melted butter

1 small onion, finely chopped

15g/^{1}/2oz/2 tbsp plain (all-purpose) flour

15ml/1 tbsp white wine vinegar

1 litre/1^3/4 pints/4 cups good chicken stock

2 egg yolks, lightly beaten

bread croûtons, fried in butter, to serve

1 Cook the garlic and onion in the oil and butter for 20 minutes, until soft.

2 Add the flour and stir to make a roux. Cook for a few minutes, then stir in the vinegar, stock and 1 litre/1^3/4 pints/ 4 cups water. Simmer for 30 minutes.

3 Before serving, whisk in the egg yolks. Put the croûtons into bowls and pour over the soup.

Energy 55kcal/229kJ; **Protein** 1.6g; **Carbohydrate** 3.6g, of which sugars 0.6g; Fat 4g, of which saturates 1.3g; **Cholesterol** 53mg; **Calcium** 13mg; **Fibre** 0.4g; **Sodium** 11mg.

Gazpacho

This traditional Spanish soup is best when sharpened with sherry vinegar.

INGREDIENTS

Serves 4

1.3–1.6kg/3–3^1/2lb ripe tomatoes, peeled

1 green (bell) pepper, seeded and chopped

2 garlic cloves, finely chopped

2 slices stale bread, crusts removed

60ml/4 tbsp extra virgin olive oil

60ml/4 tbsp sherry vinegar or wine vinegar

150ml/^{1}/4 pint/^{2}/3 cup tomato juice

300ml/^{1}/2 pint/1^1/4 cups iced water

salt and ground black pepper

For the garnishes

30ml/2 tbsp olive oil

2–3 slices stale bread, diced

1 small cucumber, peeled and finely diced

1 small onion, finely chopped

1 red and 1 green (bell) pepper, seeded and diced

2 hard-boiled eggs, chopped

1 Quarter the tomatoes and remove the cores and seeds, saving the juices.

2 Process the pepper for a few seconds in a food processor. Add the tomatoes and juices, garlic, bread, oil, vinegar and tomato juice and process. Season. Pour into a bowl, cover with clear film (plastic wrap) and chill for 12 hours.

3 To prepare the garnishes, fry the bread cubes in the olive oil for 5 minutes until golden and crisp. Drain, then place in a dish. Place each of the remaining garnishes in a separate dish. Before serving, dilute the soup with the ice-cold water. Serve with the garnishes.

Energy 376kcal/1584kJ; **Protein** 11.3g; **Carbohydrate** 38.3g, of which sugars 31.3g; **Fat** 21.1g, of which saturates 3.6g; **Cholesterol** 95mg; **Calcium** 109mg; **Fibre** 8.3g; **Sodium** 1032mg.

Hot and sour soup

The spicy flavours of this warming soup are balanced by rice vinegar. This dish makes a perfect starter to a Chinese meal.

INGREDIENTS

Serves 4

10g/¹/₄oz dried cloud ear
 (wood ear) mushrooms
8 fresh shiitake mushrooms
900ml/1¹/₂ pints/3³/₄ cups vegetable stock
75g/3oz firm tofu, cubed
50g/2oz/¹/₂ cup shredded, drained, canned
 bamboo shoots
15ml/1 tbsp caster (superfine) sugar
45ml/3 tbsp rice vinegar
15ml/1 tbsp light soy sauce
1.5ml/¹/₄ tsp chilli oil
2.5ml/¹/₂ tsp salt
large pinch of ground white pepper
15ml/1 tbsp cornflour (cornstarch)
15ml/1 tbsp cold water
1 egg white
5ml/1 tsp sesame oil
2 spring onions (scallions), cut into fine rings

1 Soak the dried cloud ears in hot water for 20 minutes until soft. Drain, trim off and discard the hard base from each cloud ear and then chop the fungus roughly.

2 Remove and discard the stems from the shiitake mushrooms. Cut the caps into thin strips.

3 Place the stock, mushrooms, tofu, bamboo shoots and cloud ear mushrooms in a large pan. Bring the stock to the boil, lower the heat and simmer for about 5 minutes.

4 Stir in the sugar, vinegar, soy sauce, chilli oil, salt and pepper. Mix the cornflour to a paste with the water. Add to the soup, stirring constantly until it thickens slightly.

5 Lightly beat the egg white, then pour it slowly into the soup in a steady stream, stirring constantly until it forms threads. Add the sesame oil, ladle into heated bowls and garnish with spring onion rings.

Energy 103kcal/429kJ; Protein 7.3g; Carbohydrate 7.3g, of which sugars 0.3g; Fat 5.1g, of which saturates 1g; Cholesterol 44mg; Calcium 135mg; Fibre 0g; Sodium 208mg.

Borscht

This beetroot soup is a traditional Eastern European dish.

INGREDIENTS

Serves 6

1 onion, chopped
1 carrot, chopped
4–6 raw or cooked, not pickled, beetroot
 (beets), 3–4 diced and 1–2 grated
400g/14oz can tomatoes
4–6 new potatoes, cut into bitesize pieces
1 small white cabbage, thinly sliced
1 litre/1³/₄ pints/4 cups vegetable stock
45ml/3 tbsp sugar
30–45ml/2–3 tbsp white wine vinegar
45ml/3 tbsp chopped fresh dill, plus extra
 to garnish
salt and ground black pepper
sour cream, to garnish

1 Put the onion, carrot, diced beetroot, tomatoes, potatoes, cabbage and stock in a pan. Bring to the boil and simmer for 30 minutes. Add the grated beetroot, sugar and vinegar and cook for 10 minutes.

2 Season. Stir the chopped dill into the soup and ladle into warmed soup bowls.

3 Place a generous spoonful of sour cream in each bowl and sprinkle some extra chopped dill over the top. Serve the soup immediately with slices of buttered rye bread.

Energy 70kcal/294kJ; Protein 2.4g; Carbohydrate 14.7g, of which sugars 9.6g; Fat 0.6g, of which saturates 0.2g; Cholesterol 0mg; Calcium 32mg; Fibre 2.8g; Sodium 43mg.

VINEGAR IN SALADS

Shaken into dressings or drizzled over to taste, all sorts of vinegars can be used to complement all varieties of salad ingredients.

Warm chorizo and spinach salad

The ingredients in this salad are tossed in olive oil and sherry or wine vinegar.

INGREDIENTS

Serves 4

90ml/6 tbsp olive oil

225g/8oz baby spinach leaves

150g/5oz chorizo sausage, very thinly sliced

30ml/2 tbsp sherry vinegar or wine vinegar

salt and ground black pepper

1 Pour the oil into a large frying pan and add the sausage. Cook gently for about 3 minutes, until the sausage slices start to shrivel slightly and colour.

2 Discard any tough stalks from the spinach. Add the spinach leaves to the sausage and remove from the heat.

3 Toss the spinach in the warm oil until it just starts to wilt. Add the sherry or wine vinegar and a little seasoning.

4 Toss the ingredients together, then serve.

Energy 300kcal/1238kJ; **Protein** 5.6g; **Carbohydrate** 4.5g, of which sugars 1.4g; **Fat** 29g, of which saturates 7g; **Cholesterol** 18mg; **Calcium** 111mg; **Fibre** 1.4g; **Sodium** 364mg.

Anchovy and roasted pepper salad

The rich balsamic vinegar in this salad really brings out the flavour of the roasted peppers.

INGREDIENTS

Serves 4

2 red, 2 orange and 2 yellow (bell) peppers, halved and seeded

50g/2oz can anchovies in olive oil

2 garlic cloves

45ml/3 tbsp balsamic vinegar

salt and ground black pepper

1 Preheat the oven to 200°C/400°F/ Gas 6. Place the peppers, cut side down, in a roasting pan. Roast for 40 minutes, until the skins are charred. Transfer the peppers to a bowl, cover with clear film (plastic wrap) and leave for 15 minutes.

2 Peel the peppers, then cut them into chunky strips. Drain the anchovies, reserving the olive oil, and halve the fillets lengthways.

3 Slice the garlic as thinly as possible and place it in a large bowl. Stir in the olive oil, vinegar and a little pepper. Add the peppers and anchovies and use a spoon and fork to fold the ingredients together. Cover and chill until ready to serve.

Energy 108kcal/453kJ; **Protein** 6g; **Carbohydrate** 16.4g, of which sugars 15.5g; **Fat** 2.4g, of which saturates 0.5g; **Cholesterol** 8mg; **Calcium** 83mg; **Fibre** 4.6g; **Sodium** 506mg.

Beetroot with fresh mint

Balsamic vinegar is mixed with oil and mint and used as a marinade.

INGREDIENTS

Serves 4

4–6 cooked beetroot (beets)

15–30ml/1–2 tbsp balsamic vinegar

30ml/2 tbsp olive oil

1 bunch fresh mint, leaves stripped and
   thinly shredded

salt and ground black pepper

1 Slice the beetroot or cut into even-size dice with a sharp knife. Put the beetroot in a small bowl.

2 Add the balsamic vinegar, olive oil and a pinch of salt to the beetroot and toss together to combine.

3 Add half of the thinly shredded fresh mint to the salad and toss together lightly until thoroughly combined. Season to taste.

4 Place in the refrigerator and chill for about 1 hour. Serve the salad garnished with the remaining shredded mint leaves sprinkled over the top.

Energy 90kcal/378kJ; **Protein** 1.7g; **Carbohydrate** 8.9g, of which sugars 8.3g; **Fat** 5.6g, of which saturates 0.8g; **Cholesterol** 0mg; **Calcium** 21mg; **Fibre** 1.9g; **Sodium** 66mg.

Sour cucumber with fresh dill

This is half pickle, half salad, and totally delicious served as a light meal or an appetizer. If possible, choose smooth-skinned, smallish cucumbers for this recipe as the larger ones tend to be less tender, with tough skins and bitter indigestible seeds.

INGREDIENTS

Serves 4

2 small cucumbers, thinly sliced

3 onions, thinly sliced

75–90ml/5–6 tbsp cider vinegar

30–45ml/2–3 tbsp chopped
   fresh dill

salt and ground black pepper

1 In a large mixing bowl, combine together the thinly sliced cucumbers and the thinly sliced onion.

2 Season the vegetables with salt and toss together until they are thoroughly combined. Leave the mixture to stand in a cool place for 5–10 minutes.

3 Add the cider vinegar, 30–45ml/2–3 tbsp water and the chopped fresh dill to the cucumber and onion mixture. Toss all the ingredients together until well combined, then chill in the refrigerator for a few hours, or until ready to serve.

Energy 89kcal/375kJ; **Protein** 2g; **Carbohydrate** 20.7g, of which sugars 18.3g; **Fat** 0.4g, of which saturates 0g; **Cholesterol** 0mg; **Calcium** 63mg; **Fibre** 2.3g; **Sodium** 9mg.

VINEGAR IN FISH AND SHELLFISH DISHES

For marinating, poaching or dressing, vinegar has many classic and contemporary associations with fish and shellfish dishes.

Japanese crab meat in vinegar

Rice vinegar is used widely in Japanese cooking – here as a dressing for crab.

INGREDIENTS

Serves 4

½ red (bell) pepper, seeded and sliced

pinch of salt

275g/10oz cooked white crab meat

about 300g/11oz cucumber, seeds removed

For the vinegar mixture

15ml/1 tbsp rice vinegar

10ml/2 tsp caster (superfine) sugar

10ml/2 tsp shoyu (soy sauce)

1 Sprinkle the pepper slices with salt and leave for 15 minutes. Rinse well and drain.

2 Combine the rice vinegar, sugar and shoyu in a bowl.

3 Loosen the crab meat and mix it with the pepper. Divide among four bowls.

4 Finely grate the cucumber. Mix the grated cucumber with the vinegar mixture, and pour a quarter into each bowl. Serve immediately.

Energy 82kcal/345kJ; **Protein** 13.3g; **Carbohydrate** 5.6g, of which sugars 5.4g; **Fat** 0.8g, of which saturates 0.1g; **Cholesterol** 50mg; **Calcium** 100mg; **Fibre** 0.9g; **Sodium** 560mg.

Seared tuna Niçoise

This simplified modern version of the traditional Tuna Niçoise is given a delicious aroma by the addition of sherry vinegar and garlic-infused oil. Serve with a green salad and fresh crusty bread.

INGREDIENTS

Serves 4

4 tuna steaks, about 150g/5oz each

45ml/3 tbsp garlic-infused olive oil

30ml/2 tbsp sherry vinegar or wine vinegar

2 eggs

salt and ground black pepper

1 Put the tuna steaks in a shallow non-metallic dish. Mix the oil and vinegar together and season with salt and pepper.

2 Pour the mixture over the tuna steaks and turn them to coat in the marinade. Cover and marinate for up to 1 hour.

3 Heat a griddle pan until smoking hot. Remove the tuna steaks from the marinade and lay them on the griddle pan. Cook for 2–3 minutes on each side, so that they are still pink in the centre. Remove from the pan and set aside.

4 Meanwhile, cook the eggs in a pan of boiling water for 5–6 minutes, then cool under cold running water. Shell the eggs and cut in half lengthways.

5 Pour the marinade on to the griddle pan and cook until it starts to bubble.

6 Divide the tuna steaks among four serving plates and top each with half an egg.

7 Drizzle the marinade over the top of the tuna and serve immediately.

Energy 578kcal/2408kJ; **Protein** 46.4g; **Carbohydrate** 15g, of which sugars 10.6g; **Fat** 37.5g, of which saturates 7.1g; **Cholesterol** 235mg; **Calcium** 127mg; **Fibre** 4.7g; **Sodium** 585mg.

Escabeche

In this classic Mexican dish, the raw fish is initially marinated in lime juice, but is then cooked before being pickled in white wine vinegar.

INGREDIENTS

Serves 4

900g/2lb whole fish fillets

juice of 2 limes

300ml/1/2 pint/11/4 cups olive oil

6 peppercorns

3 garlic cloves, sliced

2.5ml/1/2 tsp ground cumin

2.5ml/1/2 tsp dried oregano

2 bay leaves

50g/2oz/1/3 cup pickled jalapeño chilli slices, chopped

1 onion, thinly sliced

250ml/8fl oz/1 cup white wine vinegar

150g/5oz/11/4 cups green olives stuffed with pimiento, to garnish

1 Place the fish fillets in a single layer in a shallow non-metallic dish.

2 Pour the lime juice over, turn the fillets over once to ensure that they are completely coated, then cover the dish and leave to marinate for about 15 minutes.

3 Drain the fish in a colander, then pat the fillets dry with kitchen paper. Heat 60ml/4 tbsp of the oil in a large frying pan, add the fish fillets and sauté for 5–6 minutes, turning once, until they are golden brown. Use a fish slice or metal spatula to transfer them to a shallow dish that will hold them in a single layer.

4 Heat 30ml/2 tbsp of the remaining oil in a frying pan. Add the peppercorns, garlic, ground cumin, oregano, bay leaves and jalapeños, and cook over a low heat for 2 minutes, then increase the heat, add the onion slices and vinegar and bring to the boil. Lower the heat and simmer for 4 minutes.

5 Remove the pan from the heat and carefully add the remaining oil. Stir well, then pour the mixture over the fish. Leave to cool, then cover the dish and marinate for 24 hours in the refrigerator.

6 When you are ready to serve, drain off the liquid and garnish the pickled fish with the stuffed olives. Salad leaves would make a good accompaniment.

Energy 414kcal/1720kJ; **Protein** 41.9g; **Carbohydrate** 1.3g, of which sugars 1g; **Fat** 26.7g, of which saturates 3.2g; **Cholesterol** 104mg; **Calcium** 30mg; **Fibre** 0.2g; **Sodium** 137mg.

Coconut salmon

White wine vinegar forms the base of the spicy marinade for the salmon.

INGREDIENTS

Serves 4

10ml/2 tsp ground cumin

10ml/2 tsp chilli powder

2.5ml/1/2 tsp ground turmeric

30ml/2 tbsp white wine vinegar

1.5ml/1/4 tsp salt

4 salmon steaks, each about 175g/6oz

1 onion, chopped

2 fresh green chillies, seeded and chopped

2 garlic cloves, crushed

2.5cm/1in piece fresh root ginger, grated

45ml/3 tbsp vegetable oil

5ml/1 tsp ground coriander

175ml/6fl oz/3/4 cup coconut milk

1 Mix 5ml/1 tsp of the ground cumin with the chilli powder, turmeric, vinegar and salt. Rub the paste over the salmon and marinate for 15 minutes.

2 Fry the onion, chillies, garlic and ginger in the oil for 5 minutes. Put into a food processor and process to a smooth paste.

3 Return the onion paste to the pan. Add the remaining cumin, the coriander and coconut milk. Bring to the boil and simmer for 5 minutes. Add the salmon steaks, cover and cook for 15 minutes, until tender. Serve with rice.

Energy 416kcal/1729kJ; **Protein** 36.2g; **Carbohydrate** 5.2g, of which sugars 4.8g; **Fat** 27.9g, of which saturates 4.4g; **Cholesterol** 88mg; **Calcium** 75mg; **Fibre** 1.1g; **Sodium** 132mg.

VINEGAR IN POULTRY DISHES

Vinegar is very useful for marinating and dressing poultry and meat, but it is also invaluable in some recipes for braising or deglazing.

4 Grill (broil) for about 5 minutes, then turn the skewers and drizzle with the remaining oil. Grill for a further 3 minutes, or until the duck is cooked through and golden.

5 Meanwhile, melt the butter in a frying pan and cook the finely chopped shallot until softened. Add the chanterelle mushrooms and cook over a high heat for 5 minutes, stirring occasionally.

6 Poach the eggs while the chanterelles are cooking. Half fill a frying pan with water, add salt and heat until simmering. Break the eggs one at a time into a cup before tipping carefully into the water. Poach the eggs gently for 3 minutes, or until the whites are set. Use a draining spoon to transfer the eggs to a warm plate and trim off any untidy white.

7 Arrange the salad leaves on serving plates, then add the chanterelles and duck.

8 Carefully add the poached eggs. Drizzle with olive oil and season with ground black pepper.

Warm duck salad with poached eggs

This salad looks spectacular and tastes divine, and makes a perfect celebration starter or, accompanied by warm crusty bread, a light lunch or supper dish. The duck is marinated in a soy sauce and vinegar mixture.

INGREDIENTS

Serves 4

3 skinless, boneless duck breasts, thinly sliced

30ml/2 tbsp soy sauce

30ml/2 tbsp balsamic vinegar

30ml/2 tbsp groundnut (peanut) oil

25g/1oz/2 tbsp unsalted butter

1 shallot, finely chopped

115g/4oz/1½ cups chanterelle mushrooms

4 eggs

50g/2oz mixed salad leaves

salt and ground black pepper

30ml/2 tbsp extra virgin olive oil, to serve

1 Toss the duck in the soy sauce and balsamic vinegar. Cover and chill for 30 minutes to allow the duck to infuse in the soy sauce and vinegar.

2 Meanwhile, Soak 12 bamboo skewers (about 13cm/5in long) in water to prevent them from burning during cooking. Preheat the grill (broiler) to medium.

3 Thread the duck on to the skewers, pleating them neatly. Place on a grill pan and drizzle with half the oil.

Energy 271kcal/1132kJ; **Protein** 29.2g; **Carbohydrate** 1.5g, of which sugars 1.1g; **Fat** 18.6g, of which saturates 3.9g; **Cholesterol** 314mg; **Calcium** 51mg; **Fibre** 0.7g; **Sodium** 196mg.

Adobo chicken and pork cooked with vinegar and ginger

This Filipino dish can also be prepared with fish, shellfish and vegetables. Use coconut vinegar for an authentic flavour, or white wine vinegar if it is unavailable.

Sichuan chicken in kung po sauce

This is a classic Chinese dish.

INGREDIENTS

Serves 3

1 egg white

10ml/2 tsp cornflour (cornstarch)

2.5ml/^{1}/2 tsp salt

2 chicken breasts, cut into small pieces

10ml/2 tbsp yellow salted beans

15ml/1 tbsp hoisin sauce

5ml/1 tsp light brown sugar

15ml/1 tbsp rice wine

15ml/1 tbsp white wine vinegar

4 garlic cloves, crushed

150ml/^{1}/4 pint/^{2}/3 cup chicken stock

45ml/3 tbsp groundnut (peanut) oil

2–3 dried chillies, broken into small pieces

115g/4oz roasted cashew nuts

fresh coriander (cilantro) to garnish

1 Whisk the egg white, add the cornflour and salt, then stir in the chicken.

2 In a bowl, mash the beans. Stir in the hoisin sauce, sugar, rice wine, vinegar, garlic and stock.

3 In a wok, fry the chicken in the oil for 2 minutes, then drain over a bowl to collect the oil. Fry the chilli pieces in the reserved oil for 1 minute. Return the chicken to the wok with the bean sauce mixture. Bring to the boil, stir in the cashew nuts and serve garnished with coriander.

Energy 490kcal/2040kJ; **Protein** 37.7g; **Carbohydrate** 12.4g, of which sugars 2.6g; **Fat** 31.9g, of which saturates 5.6g; **Cholesterol** 82mg; **Calcium** 24mg; **Fibre** 1.9g; **Sodium** 204mg.

INGREDIENTS

Serves 4–6

30ml/2 tbsp coconut oil

6–8 garlic cloves, crushed whole

50g/2oz fresh root ginger, sliced
    into matchsticks

6 spring onions (scallions), cut into
    2.5cm/1in pieces

5–10ml/1–2 tsp whole black
    peppercorns, crushed

30ml/2 tbsp palm sugar (jaggery) or
    muscovado sugar

8–10 chicken thighs, or thighs and drumsticks

350g/12 oz pork fillet (tenderloin),
    cut into chunks

150ml/^{1}/4 pint/^{2}/3 cup coconut vinegar

150ml/^{1}/4 pint/^{2}/3 cup dark soy sauce

300ml/^{1}/2 pint/1^1/4 cups chicken stock

2–3 bay leaves

salt

stir-fried greens and cooked rice, to serve

1 Heat the oil in a wok, stir in the garlic and ginger and fry until fragrant and beginning to colour. Add the spring onions and black pepper and stir in the sugar.

2 Add the chicken and pork to the wok and fry until they begin to colour.

3 Pour in the vinegar, soy sauce and chicken stock and add the bay leaves.

4 Bring to the boil, reduce the heat, and cover. Simmer gently for 1 hour, until the meat is tender and the liquid has reduced.

5 Season the stew with salt to taste and serve with stir-fried greens and rice, over which the cooking liquid is spooned.

Energy 270kcal/1135kJ; **Protein** 42.2g; **Carbohydrate** 9g, of which sugars 7.6g; **Fat** 7.4g, of which saturates 1.6g; **Cholesterol** 118mg; **Calcium** 24mg; **Fibre** 0.6g; **Sodium** 1892mg.

VINEGAR IN MEAT DISHES

Used in marinades or cooking sauces, vinegar can be used to make meat tender and moist, with rich sweet-sour flavours as a finishing touch.

Lamb steaks with redcurrant vinegar glaze

This classic, simple dish is an excellent, quick recipe for cooking on the barbecue. The tangy flavour of redcurrants is a traditional accompaniment to lamb.

INGREDIENTS

Serves 4

4 large fresh rosemary sprigs

4 lamb leg steaks

75ml/5 tbsp redcurrant jelly

30ml/2 tbsp raspberry or red wine vinegar

salt and ground black pepper

1 Reserve the tips of the rosemary and chop the remaining leaves. Rub the chopped rosemary, salt and pepper all over the lamb.

2 Preheat the grill (broiler). Heat the redcurrant jelly gently in a small pan with 30ml/2 tbsp water. Stir in the vinegar.

3 Place the steaks on a foil-lined grill (broiler) rack and brush with a little of the redcurrant glaze. Cook for 5 minutes on each side, until deep golden, brushing with more glaze.

4 Transfer the lamb to warmed plates. Tip any juices from the foil into the remaining glaze and heat through. Pour over the lamb and serve, garnished with the reserved rosemary.

Energy 301kcal/1,258kJ; **Protein** 24g; **Carbohydrate** 12g, of which sugars 12g; **Fat** 17g, of which saturates 8g; **Cholesterol** 94mg; **Calcium** 10mg; **Fibre** 0.0g; **Sodium** 100mg.

Sweet-and-sour lamb

Tender lamb chops are marinated, then cooked in balsamic vinegar.

INGREDIENTS

Serves 4

8 French-trimmed lamb loin chops

90ml/6 tbsp balsamic vinegar

30ml/2 tbsp caster (superfine) sugar

30ml/2 tbsp olive oil

salt and ground black pepper

1 Put the lamb chops in a shallow, non-metallic dish and drizzle over the balsamic vinegar. Sprinkle with sugar and season with salt and black pepper. Turn the chops to coat, cover with clear film (plastic wrap) and marinate for 20 minutes.

2 Heat the oil in a large frying pan and add the chops, reserving the marinade. Cook for 3–4 minutes on each side.

3 Pour the marinade into the pan and leave to bubble for about 2 minutes, or until reduced slightly. Remove from the pan and serve immediately.

Energy 258kcal/1077kJ; **Protein** 19.6g; **Carbohydrate** 7.9g, of which sugars 7.9g; **Fat** 16.7g, of which saturates 6g; **Cholesterol** 76mg; **Calcium** 12mg; **Fibre** 0g; **Sodium** 87mg.

Beef with Asian pear

Rice vinegar is used to marinate the pear.

INGREDIENTS

Serves 4

300g/11oz fillet steak, thinly sliced

15ml/1 tbsp dark soy sauce

1 garlic clove, crushed

10ml/2 tsp sesame oil

1 Asian pear

15ml/1 tbsp sugar

15ml/1 tbsp rice vinegar

vegetable oil, for cooking

30ml/2 tbsp pine nuts

5 cooked chestnuts, finely chopped

salt and ground black pepper

1 Mix the steak with the soy sauce, garlic, sesame oil and a little salt and pepper and marinate for 20 minutes.

2 Meanwhile, peel, core and slice the pear, then cut the slices into fine strips. Place in a small bowl and pour in cold water to cover. Stir the sugar and rice vinegar into the bowl. Leave to stand for 5 minutes, then drain and set aside.

3 Heat a frying pan over a high heat and add a little vegetable oil. Add the beef with its marinade and sauté briefly, then reduce the heat and fry gently until the meat is well cooked. Transfer to a serving dish. Add the pear, pine nuts and chestnuts, toss together and serve.

Energy 234kcal/976kJ; **Protein** 18.6g; **Carbohydrate** 8.9g, of which sugars 5.2g; **Fat** 14g, of which saturates 3.5g; **Cholesterol** 44mg; **Calcium** 15mg; **Fibre** 1.5g; **Sodium** 300mg.

Steak with warm tomato salsa

A refreshing, tangy salsa of tomatoes, spring onions and balsamic vinegar makes a colourful topping for chunky, pan-fried steaks. Serve with potato wedges and a mixed leaf salad with a mustard dressing.

INGREDIENTS

Serves 4

2 steaks, about 2cm/3/4 in thick

4 large plum tomatoes

2 spring onions (scallions)

30ml/2 tbsp balsamic vinegar

salt and ground black pepper

1 Trim any excess fat from the steaks, then season on both sides with salt and pepper.

2 Heat a non-stick frying pan and cook the steaks for 3 minutes on each side for medium rare, or longer if you prefer.

3 Meanwhile, put the tomatoes in a heatproof bowl, cover with boiling water and leave for 1–2 minutes, until the skins start to split.

4 Drain and peel the tomatoes, then halve them and scoop out the seeds. Dice the tomato flesh. Thinly slice the spring onions.

5 Transfer the steaks to plates and keep warm.

6 Add the vegetables, balsamic vinegar, 30ml/2 tbsp water and a little seasoning to the cooking juices in the pan and stir briefly until warm, scraping up any meat residue. Spoon the salsa over the steaks to serve.

Energy 291kcal/1215kJ; **Protein** 35.3g; **Carbohydrate** 5g, of which sugars 5g; **Fat** 14.5g, of which saturates 5.9g; **Cholesterol** 87mg; **Calcium** 22mg; **Fibre** 1.7g; **Sodium** 110mg.

VINEGAR IN BAKING

It may seem surprising, but vinegar is a useful ingredient for baking many sweet dishes. Use to balance the sweetness of tarts and ensure cakes rise beautifully.

2 Mix the bicarbonate of soda with the vinegar and, as it froths, quickly stir it into the mixture. Cover the bowl and leave at room temperature for 8 hours.

3 Preheat the oven to 180°C/360°F/ Gas 4. Grease a shallow 23cm/9 in round cake tin (pan) and line its base with baking parchment. Spoon the mixture into the prepared tin and level the top.

Overnight cake

Baking powder is a mixture of an acid and an alkali. When combined with moisture and heated, this mixture gives off carbon dioxide, which creates bubbles in a mixture and makes it rise. Cooking sets the mixture and traps the bubbles. Using vinegar, an acid, in the same mixture as bicarbonate of soda, an alkali, has the same effect. The following is a traditional British recipe for an eggless fruit cake. The vinegar will lose its acidity once the cake is cooked, providing just a hint of sharpness.

INGREDIENTS

Serves 4

225g/8oz/2 cups plain (all-purpose) flour
5ml/1 tsp ground cinnamon
5ml/1 tsp ground ginger
115g/4oz/^{1}/2 cup butter, diced
115g/4oz/^{2}/3 cup mixed dried fruit
300ml/^{1}/2 pint/1^1/4 cups milk
2.5ml/1/2 tsp bicarbonate of soda (baking soda)
15ml/1 tbsp cider vinegar or white vinegar

1 Sift the flour and spices together. Add the butter and rub in until the mixture resembles fine breadcrumbs. Stir in the dried fruit and enough milk to make a soft mix.

4 Put into the oven and cook for 1 hour or until firm to the touch and cooked through – a skewer inserted in the centre should come out clean. If the top starts to get too brown, cover with baking parchment.

5 Cool in the tin for 20 minutes, turn out and cool completely on a wire rack.

Energy 2069kcal/8681kJ; **Protein** 34.7g; **Carbohydrate** 267.9g, of which sugars 96.5g; **Fat** 103g, of which saturates 63.6g; **Cholesterol** 263mg; **Calcium** 780mg; **Fibre** 9.5g; **Sodium** 888mg.

Pavlova

There are three main elements that make pavlova different from other meringues: the addition of cornflour and vinegar; folding in, rather than whisking in, the sugar; and the depth of the cooked meringue. These differences help to create the soft, chewy centre.

INGREDIENTS

Serves 4–6

4 egg whites

220g/7oz sugar

15–30ml/1–2 tbsp cornflour (cornstarch)

15ml/1 tbsp white wine vinegar

300ml/1/2 pint/1 1/2 cups whipped cream

450g/1lb/4 cups mixed berries

1 Whisk the egg whites to form soft peaks. Whisk in most of the sugar, reserving a small amount. Sift the cornflour over the meringue, then fold in the remaining sugar. Add the vinegar.

2 Fold in the vinegar with a large metal spoon until blended. Work gently to avoid knocking the air from the meringue.

3 Preheat the oven to 140°C/275°F/Gas 1. Prepare a baking sheet, drawing a 23cm/9in circle in heavy pencil on the reverse of the paper.

4 Spread half the mixture into a thick, flat neat round, then spoon the rest in high swirls around the edge to create a border. Bake for 1–1 1/2 hours until the meringue is firm, checking frequently.

5 When cooked and cooled, peel off the paper and fill the pavlova shell with the whipped cream and the berries.

Energy 565kcal/2368kJ; **Protein** 5.6g; **Carbohydrate** 73.5g, of which sugars 66.6g; **Fat** 29.7g, of which saturates 18.5g; **Cholesterol** 78.8mg; **Calcium** 96.5mg; **Fibre** 1.3g; **Sodium** 104.8mg.

Currant and walnut tart

Add a little white wine vinegar to this tart to counter the sweetness.

INGREDIENTS

Serves 4

250g/9oz sweet pastry

1 egg

75g/3oz/scant 1/2 cup soft light brown sugar

50g/2oz/1/4 cup butter, melted

10ml/2 tsp white wine vinegar

115g/4oz/1/2 cup currants

25g/1oz/1/4 cup chopped walnuts

double (heavy) cream, to serve (optional)

1 Line a 20cm/8in flan tin (tart pan) with sweet pastry. Preheat the oven to 190°C/375°F/Gas 5.

2 Mix the egg, sugar and melted butter together. Stir the vinegar, currants and walnuts into the mixture.

3 Pour the mixture into the pastry case and bake for 30 minutes. Remove from the oven when thoroughly cooked, take out of the flan tin and leave to cool on a wire rack for at least 30 minutes.

4 Serve the tart on its own or with a dollop of fresh cream, if using.

Energy 312kcal/1,307kJ; **Protein** 3.4g; **Carbohydrate** 41.1g, of which sugars 41g; **Fat** 16.1g, of which saturates 7.3g; **Cholesterol** 74mg; **Calcium** 54mg; **Fibre** 0.8g; **Sodium** 99mg.

VINEGAR IN DRINKS

A little vinegar in sparkling mineral water makes a refreshing drink and an excellent alternative to an alcoholic drink or glass of wine. It is also used to give an edge to some cocktails.

Virgin prairie oyster

This delicious non-alcoholic version of the classic hangover cure will help you to avoid hangovers altogether.

INGREDIENTS

Makes about 200ml/7fl oz/scant 1 cup

175ml/6fl oz tomato juice

1/2 measure/2 tsp Worcestershire sauce

1/4–1/2 measure/1–2 tsp balsamic vinegar

1 egg yolk

cayenne pepper, to taste

1 Measure the tomato juice into a large bar glass and stir over plenty of ice until well chilled.

2 Strain into a tall tumbler half-filled with ice cubes.

3 Add the Worcestershire sauce and balsamic vinegar to taste and mix with a swizzle-stick.

4 Carefully float the egg yolk on top of the drink and lightly dust with cayenne pepper.

Energy 92kcal/388kJ; **Protein** 4.4g; **Carbohydrate** 6.8g, of which sugars 6.7g; **Fat** 5.5g, of which saturates 1.6g; **Cholesterol** 202mg; **Calcium** 60mg; **Fibre** 1.1g; **Sodium** 532mg.

Prairie oyster

This cocktail, containing a dash of white wine vinegar, is said to cure hangovers.

INGREDIENTS

Makes about 45ml/3 tbsp

1 measure/1¹/2 tbsp sweet (malmsey) Madeira, or cognac

1/2 measure/1 tsp white wine vinegar

1/2 measure/1 tsp Worcestershire sauce

pinch cayenne pepper

dash Tabasco

1 egg yolk

1 Add the Madeira or cognac, white wine vinegar, Worcestershire sauce, cayenne pepper and Tabasco to a small tumbler.

2 Mix well without ice, and then spoon the yolk very gently on top.

3 The preparation should then be swallowed in one gulp, without breaking the egg yolk.

Energy 112kcal/463kJ; **Protein** 3g; **Carbohydrate** 0.8g, of which sugars 0.7g; **Fat** 5.5g, of which saturates 1.6g; **Cholesterol** 202mg; **Calcium** 33mg; **Fibre** 0g; **Sodium** 69mg.

Vinegar spritzers

Mixing a little vinegar with mineral water makes a surprisingly tasty summer drink.

INGREDIENTS

Makes one glass
10ml/2 tsp fruit vinegar, such as raspberry,
  fig or blackberry
slices of cucumber, orange and apple
sprig of fresh mint
sparkling mineral water

1 Pour the fruit vinegar into a glass.

2 Add the slices of cucumber, orange and apple with a sprig of fresh mint and top up with sparkling mineral water.

Variations

Place 10ml/2 tsp raspberry vinegar in a glass and top up with sparkling water.

Sweet moscatel vinegar is fabulous in mineral water. Pour about 10ml/2 tsp in a wine glass and top up with still water.

Add balsamic vinegar to sparkling mineral water, allowing about 5ml/ 1 tsp per glass. This is similar to water with Angostura bitters.

Energy 24kcal/104kJ; **Protein** 0.3g; **Carbohydrate** 6.1g, of which sugars 6.1g; **Fat** 0.1g, of which saturates 0g; **Cholesterol** 0mg; **Calcium** 5mg; **Fibre** 1.1g; **Sodium** 4mg.

Vinegar smoothies

Adding vinegar to fruit smoothies gives them a tangy kick.

INGREDIENTS

Makes about 400ml/³/4 pint/scant 2 cups
15ml/1 tbsp fruit vinegar, such as
  raspberry, orange or blackberry
300ml/¹/2 pint/1¹/4 cup natural (plain) yogurt
fruit juice, such as apple

1 Pour the yogurt into a tall glass and add the fruit vinegar.

2 Top up the glass with fruit juice.

Variations

Try adding 10ml/2 tsp fermented fig vinegar for a rich, sweet-sour, caramel-fig flavoured drink.

Fruit vinegars such as raspberry, strawberry or blackberry are very good with banana or apple in smoothies, adding a pleasing sharpness.

Combine raspberry vinegar, honey and yogurt for a invigorating smoothie. Stir in a little raspberry and cranberry juice to thin the drink if necessary.

Energy 204kcal/861kJ; **Protein** 15.7g; **Carbohydrate** 31.5g, of which sugars 31.5g; **Fat** 3.1g, of which saturates 1.5g; **Cholesterol** 4mg; **Calcium** 576mg; **Fibre** 1.6g; **Sodium** 254mg.

Sparkling elderflower drink

This refreshing drink will keep in airtight bottles for up to a month.

INGREDIENTS

Makes 9.5 litres/2 gallons/2.4 US gallons
24 elderflower heads
2 lemons
1.3kg/3lb/scant 7 cups granulated
  (white) sugar
30ml/2 tbsp white wine vinegar or
  cider vinegar
9 litres/2 gallons/2.4 US gallons water

1 Find a clean bucket that you can cover with a clean cloth. Put all the ingredients into it.

2 Cover the bucket with a plastic sheet and leave overnight.

3 Strain the elderflower drink then pour into bottles, leaving a space of about 2.5cm/1in at the top.

4 Seal the bottles and leave for 2 weeks in a cool place. A fermentation takes place and the result is a delightful, sparkling, refreshing, non-alcoholic drink. Serve chilled with slices of lemon.

Energy 186kcal/793kJ; **Protein** 7.5g; **Carbohydrate** 34.6g, of which sugars 16.4g; **Fat** 3.1g, of which saturates 0.4g; **Cholesterol** 1mg; **Calcium** 137mg; **Fibre** 5.5g; **Sodium** 51mg.

VINEGAR IN PICKLES

Pickles can be sharp or sweet or a combination of the two. They are usually made by preserving raw or lightly cooked fruit or vegetables in spiced vinegar.

Shallots in balsamic vinegar

Whole shallots cooked in balsamic vinegar and herbs have a gentle taste.

Pickled onions

These powerful pickles should be made with malt vinegar and stored for at least 6 weeks before eating.

INGREDIENTS

Makes about 4 jars

1kg/2 1/4lb pickling onions, peeled

115g/4oz/1/2 cup salt

750ml/1 1/4 pints/3 cups malt vinegar

15ml/1 tbsp sugar

2–3 dried red chillies

5ml/1 tsp brown mustard seeds

15ml/1 tbsp coriander seeds

5ml/1 tsp allspice berries

5ml/1 tsp black peppercorns

5cm/2in piece fresh root ginger, sliced

2–3 blades mace

2–3 fresh bay leaves

1 Place the peeled onions in a bowl, cover with cold water, then drain the water into a pan.

2 Add the salt and heat slightly to dissolve, then cool before pouring the brine over the onions.

3 Place a plate inside the top of the bowl and weigh it down slightly so that it keeps all the onions submerged in the brine. Leave to stand for 24 hours.

4 Meanwhile, place the vinegar in a large pan. Wrap all the remaining ingredients, except the bay leaves, in a piece of muslin (cheesecloth). Bring to the boil, simmer for about 5 minutes, then remove the pan from the heat. Set aside and leave to infuse overnight.

5 The next day, drain the onions, rinse and pat dry. Pack them into sterilized 450g/1lb jars. Add some or all of the spice from the vinegar, except the ginger. The pickle will become hotter if you add the chillies. Pour the vinegar over to cover and add the bay leaves.

6 Seal the jars with non-metallic lids and store in a cool, dark place for at least 6 weeks before eating.

Energy 669kcal/2,775kJ; **Protein** 48.7g; **Carbohydrate** 5g, of which sugars 3.8g; **Fat** 45.9g, of which saturates 11.1g; **Cholesterol** 250mg; **Calcium** 37mg; **Fibre** 0.9g; **Sodium** 196mg.

INGREDIENTS

Makes 1 large jar

500g/1 1/4lb shallots

30ml/2 tbsp muscovado (molasses) sugar

fresh thyme sprigs

300ml/1/2 pint/1 1/4 cups balsamic vinegar

1 Put the unpeeled shallots in a bowl. Pour over boiling water and leave to stand for 2 minutes to loosen the skins. Drain and peel the shallots, leaving whole.

2 Put the sugar, thyme and vinegar in a pan and bring to the boil.

3 Add the shallots, cover and simmer gently for 40 minutes, or until the shallots are just tender.

4 Transfer to a warmed sterilized jar, packing the shallots down well. Seal and label the jar, then store in a cool, dark place for about 1 month before eating.

Energy 298kcal/1254kJ; **Protein** 6.1g; **Carbohydrate** 70.9g, of which sugars 59.3g; **Fat** 1g, of which saturates 0g; **Cholesterol** 0mg; **Calcium** 141mg; **Fibre** 7g; **Sodium** 17mg.

Pickled red cabbage

This delicately spiced and vibrant pickle is made with red wine vinegar.

INGREDIENTS

Serves 4

675g/1¹/₂lb/6 cups red cabbage, shredded

1 large Spanish onion, sliced

30ml/2 tbsp sea salt

600ml/1 pint/2¹/₂ cups red wine vinegar

75g/3oz/6 tbsp light muscovado
  (brown) sugar

15ml/1 tbsp coriander seeds

3 cloves

2.5cm/1in piece fresh root ginger

1 whole star anise

2 bay leaves

4 eating apples

1 Put the cabbage and onion in a bowl, add the salt and mix well. Transfer the mixture into a colander over a bowl and leave to drain overnight.

2 The next day, rinse the vegetables, drain and pat dry using kitchen paper. Pour the vinegar into a pan, add the sugar, spices and bay leaves and bring to the boil. Remove from the heat and cool.

3 Core and chop the apples, then layer with the cabbage and onions in sterilized preserving jars. Pour over the cooled spiced vinegar. Seal and store for 1 week before eating. Eat within 2 months. Once opened, keep chilled.

Energy 674kcal/2868kJ; **Protein** 12g; **Carbohydrate** 161.4g, of which sugars 159.3g; **Fat** 2g, of which saturates 0g; **Cholesterol** 0mg; **Calcium** 405mg; **Fibre** 23g; **Sodium** 64mg.

Pickled ginger

Warming, good for the heart, and believed to aid digestion, ginger finds its way into oriental cooking in many dishes from salads, soups and stir-fries to puddings. Chinese in origin, ginger pickled in rice vinegar is often served as a condiment with broths, noodles and rice.

INGREDIENTS

Serves 4–6

225g/8oz fresh young ginger, peeled

10ml/2 tsp salt

200ml/7fl oz/1 cup white rice vinegar

50g/2oz/¹/₄ cup sugar

1 Place the ginger in a bowl and sprinkle with salt. Cover and place in the refrigerator for at least 24 hours.

2 Drain off any excess liquid and pat the ginger dry with a clean dish towel. Slice each knob of ginger very finely along the grain, like thin rose petals, and place them in a clean bowl or a sterilized jar suitable for storing.

3 In a small bowl, beat the vinegar and 50ml/2fl oz/1/4 cup cold water with the sugar, until it has all completely dissolved.

4 Pour the pickling liquid over the ginger and cover or seal. Store in the refrigerator or a cool place for about 1 week.

Energy 36kcal/151kJ; **Protein** 0.2g; **Carbohydrate** 9.1g, of which sugars 9.1g; **Fat** 0.1g, of which saturates 0g; **Cholesterol** 0mg; **Calcium** 20mg; **Fibre** 0.4g; **Sodium** 678mg.

VINEGAR IN RELISHES AND CHUTNEYS

Chutneys and relishes are made from finely cut ingredients, cooked with vinegar, a sweetener and frequently spices to make a thick, savoury jam-like mixture.

Tart tomato relish

Adding lime to this relish gives it a wonderfully tart, tangy flavour.

INGREDIENTS

Makes about 500g/1¹/₄lb

2 pieces preserved stem ginger

1 lime

450g/1lb cherry tomatoes

115g/4oz/¹/₂ cup muscovado (molasses) sugar

120ml/4fl oz/¹/₂ cup white wine vinegar

5ml/1 tsp salt

1 Coarsely chop the preserved stem ginger. Slice the lime thinly, including the rind, then chop into small pieces. Place the cherry tomatoes, sugar, vinegar, salt, ginger and lime in a large heavy pan.

2 Bring to the boil, stirring until the sugar dissolves. Simmer rapidly for 45 minutes. Stir until the liquid has evaporated and the relish is thick and pulpy.

3 Leave to cool for about 5 minutes, then spoon into sterilized jars. Leave to cool, then cover and store in the refrigerator for up to 1 month.

Energy 530kcal/2262kJ; **Protein** 3.7g; **Carbohydrate** 134.1g, of which sugars 134.1g; **Fat** 1.4g, of which saturates 0.5g; **Cholesterol** 0mg; **Calcium** 93mg; **Fibre** 4.5g; **Sodium** 2012mg.

Cranberry and red onion relish

This sophisticated relish mixes the richness of caramelized onions with the tartness of cranberries and vinegar.

INGREDIENTS

Makes about 900g/2lb

30ml/2 tbsp olive oil

450g/1lb small red onions, finely sliced

225g/8oz/1 cup soft light brown sugar

450g/1lb/4 cups fresh or frozen cranberries

120ml/4fl oz/¹/₂ cup red wine vinegar

120ml/4fl oz/¹/₂ cup red wine

15ml/1 tbsp mustard seeds

2.5ml/¹/₂ tsp ground ginger

30ml/2 tbsp orange liqueur or port

salt and ground black pepper

1 Heat the olive oil in a large pan, add the sliced onions and cook over a low heat for about 15 minutes, stirring occasionally, until they have softened.

2 Add 30ml/2 tbsp of the sugar and cook for a further 5 minutes, or until the onions are caramelized.

3 Meanwhile, put the cranberries in a pan with the remaining sugar, and the vinegar, wine, mustard seeds and ginger. Heat gently until the sugar has dissolved, then cover and bring to the boil. It is important to cover the pan when cooking the cranberries because they can pop out of the pan during cooking and are very hot.

4 Simmer for 15 minutes, until the berries have burst and are tender, then stir in the caramelized onions. Increase the heat slightly and cook uncovered for a further 10 minutes, stirring frequently until well reduced and thickened. Remove from the heat, then season to taste.

5 Transfer to warmed sterilized jars. Spoon a little of the orange liqueur or port over the top of each, then cover and seal. Store the jars in a cool place for up to 6 months. Once opened, store in the refrigerator and then use within 1 month.

Energy 1532kcal/6486kJ; **Protein** 8g; **Carbohydrate** 314.6g, of which sugars 304.2g; **Fat** 23.3g, of which saturates 3.2g; **Cholesterol** 0mg; **Calcium** 259mg; **Fibre** 13.5g; **Sodium** 46mg.

Chunky pear and walnut chutney

The perfect balance of cider vinegar and tart apples and pears makes this a surprisingly mellow accompaniment to cheese as well as grains including pilaffs and tabbouleh.

INGREDIENTS

Makes about 1.8kg/4lb

1.2kg/2¹/²lb firm pears
225g/8oz tart cooking apples
225g/8oz onions
450ml/³/4 pint/scant 2 cups cider vinegar
175g/6oz/generous 1 cup sultanas
   (golden raisins)
finely grated rind and juice of 1 orange
400g/14oz/2 cups sugar
115g/4oz/1 cup walnuts, roughly chopped
2.5ml/¹/2 tsp ground cinnamon

1 Peel and core the pears and apples, then chop them into 2.5cm/1in chunks. Peel and quarter the onions, then chop into pieces of the same size.

2 Place in a preserving pan with the vinegar. Slowly bring to the boil, then reduce the heat and simmer for 40 minutes, until the apples, pears and onions are tender, stirring occasionally.

3 Meanwhile, put the sultanas in a bowl, pour over the orange juice and leave to soak.

4 Add the sugar, sultanas, and orange rind and juice to the pan.

5 Gently heat the mixture until the sugar has dissolved, then simmer for 30–40 minutes, or until the chutney is thick and no excess liquid remains. Stir frequently towards the end of cooking to prevent the chutney sticking on the bottom of the pan.

6 Gently toast the walnuts in a non-stick pan over a low heat, stirring constantly, for 5 minutes, until golden. Stir the nuts into the chutney with the cinnamon.

7 Spoon the chutney into warmed sterilized jars, cover and seal. Store in a cool, dark place, then leave to mature for at least 1 month. Use within 1 year. Once opened, store in the refrigerator and use within 2 months.

Energy 3501kcal/14,797kJ; **Protein** 29.8g; **Carbohydrate** 705.3g, of which sugars 699.3g; **Fat** 81.4g, of which saturates 6.4g; **Cholesterol** 0mg; **Calcium** 603mg; **Fibre** 40.7g; **Sodium** 189mg.

Kashmir chutney

This sweet, chunky, spicy chutney is perfect served with cold meats.

INGREDIENTS

Makes about 2.75kg/6lb

1kg/2¹/4lb green eating apples
15g/¹/2oz garlic cloves
1 litre/1³/4 pints/4 cups malt vinegar
450g/1lb dates
115g/4oz preserved stem ginger
450g/1lb/3 cups raisins
450g/1lb/2 cups soft light brown sugar
2.5ml/¹/2 tsp cayenne pepper
30ml/2 tbsp salt

1 Quarter the apples, remove the cores and chop coarsely. Peel and chop the garlic.

2 Place the apple with the chopped garlic in a pan with enough vinegar to cover. Bring to the boil and boil for 10 minutes.

3 Chop the dates and ginger and add them to the pan, with the rest of the ingredients. Cook gently for 45 minutes.

4 Spoon the mixture into warmed sterilized jars and seal immediately. Store in a cool, dark place and use within 1 year. Once opened, store in the refrigerator and use within 2 months.

Energy 3920kcal/16,737kJ; **Protein** 22.6g; **Carbohydrate** 1014.4g, of which sugars 1012.2g; **Fat** 3.3g, of which saturates 0g; **Cholesterol** 0mg; **Calcium** 599mg; **Fibre** 33.7g; **Sodium** 12139mg.

VINEGAR IN JELLIES

Savoury jellies are classic accompaniments for roasted meats. Fruit or vegetables are cooked in vinegar and boiled with sugar to setting point.

Plum and apple jelly

Serve this rich jelly with roast meats.

INGREDIENTS

Makes about 1.3kg/3lb

900g/2lb plums, stoned (pitted) and chopped

450g/1lb tart cooking apples, chopped

150ml/^{1}/$_4$ pint/^{2}/$_3$ cup cider vinegar

750ml/1^1/$_4$ pints/3 cups water

675g/1^1/$_2$lb/scant 3^1/$_2$ cups sugar

Red pepper and chilli jelly

The hint of chilli in this jelly makes it ideal for spicing up sausages or burgers.

INGREDIENTS

Makes about 900g/2lb

8 red (bell) peppers, quartered and seeded

4 fresh red chillies, halved and seeded

1 onion, roughly chopped

2 garlic cloves, roughly chopped

250ml/8fl oz/1 cup water

250ml/8fl oz/1 cup white wine vinegar

7.5ml/1^1/$_2$ tsp salt

450g/1lb/2^1/$_4$ cups sugar

25ml/1^1/$_2$ tbsp powdered pectin

1 Arrange the quartered red peppers, skin side up, on a rack in a grill (broiling) pan and grill (broil) until the skins blacken and blister.

2 Put the peppers in a polythene bag to steam for about 10 minutes, then carefully remove the skins.

3 Put the peppers, chillies, onion, garlic and water in a food processor and process to a purée. Press through a sieve (strainer) set over a bowl, pressing with a spoon to extract as much juice as possible. There should be about 750ml/1^1/$_4$ pints/3 cups.

4 Scrape the purée into a large pan, then stir in the vinegar and salt. Combine the warmed sugar and pectin, then stir it into the puréed pepper mixture. Heat gently, stirring, until the sugar and pectin have dissolved, then bring to the boil.

5 Cook the jelly, stirring, for 4 minutes, then remove the pan from the heat. Pour into warmed, sterilized jars. Leave to cool and set, then cover, label and store in a cool dark place. Use within 1 year. Once opened, store in the refrigerator and use within 2 months.

Energy 2275kcal/9665kJ; **Protein** 18g; **Carbohydrate** 571g, of which sugars 565.1g; **Fat** 6.1g, of which saturates 1.5g; **Cholesterol** 0mg; **Calcium** 373mg; **Fibre** 24.8g; **Sodium** 89mg.

1 In a pan, bring the fruit, vinegar and water to the boil, reduce the heat, cover and simmer for 30 minutes. Pour into a sterilized jelly bag suspended over a bowl.

2 Drain for 3 hours. Measure the juice into a pan, adding 450g/1lb/2^1/$_4$ cups sugar for every 600ml/1 pint/2^1/$_2$ cups juice. Bring to the boil, stirring, until the sugar dissolves. Boil for 10 minutes, or to setting point (105°C/220°F). Remove from heat and skim off scum. Pour into warmed sterilized jars. Cover and seal.

3 Store in a cool, dark place and use within 2 years. Once opened, keep refrigerated and use within 2 months.

Energy 2803kcal/11,963kJ; **Protein** 5.5g; **Carbohydrate** 740.7g, of which sugars 740.7g; **Fat** 0.4g, of which saturates 0g; **Cholesterol** 0mg; **Calcium** 401mg; **Fibre** 6.4g; **Sodium** 49mg.

Minted gooseberry jelly

This classic, tart jelly takes on a pinkish tinge during cooking, not green as one would expect.

INGREDIENTS

Makes about 1.2kg/2¹/₂lb

1.3kg/3lb/12 cups gooseberries

1 bunch fresh mint

750ml/1¹/₄ pints/3 cups cold water

400ml/14fl oz/1 ²/₃ cups white wine vinegar

about 900g/2lb/4¹/₂ cups preserving or
  granulated (white) sugar

45ml/3 tbsp chopped fresh mint

1 Place the gooseberries, mint and water in a preserving pan. Bring to the boil, reduce the heat, cover and simmer for 30 minutes, until the gooseberries are soft. Add the vinegar and simmer, uncovered, for a further 10 minutes.

2 Pour the fruit and juices into a sterilized jelly bag suspended over a bowl. Leave to drain for at least 3 hours, until the juices stop dripping, then measure the strained juices back into the cleaned preserving pan.

3 Add 450g/1lb/2¹/₄ cups sugar for every 600ml/1 pint/2¹/₂ cups juice, then heat gently, stirring, until dissolved. Bring to the boil. Cook for 15 minutes to setting point (105°C/220°F). Remove from the heat and skim off any scum. Cool until a thin skin forms, then stir in the mint.

4 Pour into sterilized jars, cover and seal. Use within 1 year. Once opened, keep chilled. Use within 3 months.

Energy 3641kcal/15,534kJ; **Protein** 10g;
Carbohydrate 955.5g, of which sugars 955.5g; **Fat** 2g,
of which saturates 0g; **Cholesterol** 0mg; **Calcium**
617mg; **Fibre** 12g; **Sodium** 64mg.

Tomato and herb jelly

This dark golden jelly is delicious served with grilled meats, especially lamb.

INGREDIENTS

Makes about 1.3kg/3lb

1.8kg/4lb tomatoes, washed and quartered

2 lemons, washed and chopped into pieces

2 bay leaves

250ml/8 fl oz/1 cup malt vinegar

300ml/¹/₂ pint/1¹/₄ cups cold water

bunch of fresh rosemary

about 900g/2lb/4¹/₂ cups sugar

1 Put the tomatoes and lemons in a large pan with the bay leaves and add the vinegar and water. Add the rosemary, bring to the boil, then reduce the heat. Cover and simmer for 40 minutes, until the tomatoes are soft.

2 Pour into a sterilized jelly bag suspended over a bowl. Drain for 3 hours. Measure the juice into a clean pan, adding 450g/1lb/2¹/₂ cups juice. Heat gently, stirring, until the sugar dissolves. Boil for 10 minutes, to setting point (105°C/220°F), then remove from the heat. Skim off any scum. Leave for a few minutes until a skin forms.

3 Pour the jelly into sterilized jars. Cover and seal when cold. Store in a cool, dark place and use within 1 year. Once opened, keep chilled. Use within 3 months.

Energy 3767kcal/16078kJ; **Protein** 13.6g;
Carbohydrate 980.8g, of which sugars 980.8g;
Fat 3.9g, of which saturates 1.3g; **Cholesterol** 0mg;
Calcium 568mg; **Fibre** 13g; **Sodium** 171mg.

OIL

Since ancient times, different oils have
been used as a vital source of light, fuel
and food, as well as featuring in religious
rites and rituals. In today's households
a variety of oils still have an important
part to play, and are used not only in
cooking but for health and diet, in skincare
and for wellbeing and indulgence.

*Above: Oils are naturally moisturizing and have been used with herbs
for skincare since the days of the ancient Egyptian pharaohs.*
Left: Olive oil is an essential ingredient in Mediterranean cooking.

ANCIENT OIL

From beautifying to embalming, as currency for rulers or chrism for religious leaders, oil has always held physical, social and spiritual significance.

Olive oil was used by the ancient Greeks, Egyptians and Romans, but other oils have been around just as long – nut oils such as groundnut (peanut) oil were used by the Aztecs, while sesame oil was used throughout Asia.

The first production of olive oil

Wall paintings and commercial records show that olives grew in ancient Egypt, and offerings of olive branches – such as those found in the tomb of Tutankhamun – accompanied the dead. The Egyptians used olive oil extensively, for cooking and lighting, in medicine and ritual. They ate the olives they grew, but for the highest quality oil they depended on imports from Palestine, Syria and Crete. The Cretans were producing oil in the third millennium BC and it is possible that this profitable trade was the source of the great wealth of the Minoans.

Above: This 3rd-century Roman mosaic shows men pressing olives to extract oil.

Above: These giant amphoras from the palace at Knossos were used to store oil.

Inhabited during the neolithic period, from the 6th century BC, probably by people who came from Asia Minor, the remains at Knossos on Crete reveal an amazing civilization. A British archaeologist, Arthur Evans, led the excavation of the site in the early 20th century and introduced the term Minoan for the civilization after the ruler King Minos. The discoveries exposed a wealth of information on the architecture, lifestyles and trading of the Minoan civilization, including underground storerooms, jars and vessels that were used for wine and oil. Oil-burning lamps lit the corridors and it was probably one of these that ignited the blaze that finally destroyed the palace. Remains of mills as well as separation tanks and storage vessels provided evidence for the early olive oil production on Crete, from where it was exported to Aegean islands and mainland Greece.

By the 6th century BC, the Greeks had become major exporters of oil, and the olive's status was so high that olive groves were regarded as sacred. In a curious pre-figuring of the modern idea of 'virgin' olive oil, only the chaste were allowed to look after the trees and harvest the crop, and this custom survived into the Middle Ages. At the Panathenaic Games in Athens, held in honour of Athene, the athletes competed for amphorae decorated with paintings of the goddess and filled with olive oil.

Olive oil was vital to the wealth as well as the welfare of ancient Greek and Phoenician civilizations, who traded it and thus spread the appreciation and production from the Eastern Mediterranean through to North Africa, Italy, Spain and Provence in France.

Olive oil was of great importance to the Roman Empire. The Romans ate olives, and developed new methods of curing them, but their greatest contribution to oil production was the invention of the screw press. This enabled them to crush the fruit mechanically to extract the maximum amount of oil from the flesh, but could be set to avoid breaking the stones (pits) and contaminating the oil.

On their travels, the Romans often demanded payment in olive oil. The expansion of the Roman Empire increased the demand for olive oil, and Romans planted trees or traded oil in all the lands they conquered, regarding those who ate animal fats rather than olive oil as barbarians. Olive groves flourished throughout Italy, in southern France and Spain, on all the islands of the Mediterranean and along the north African coast from Tripoli to Algeria.

Other oils

Cultivation of almonds is as ancient as that of olives but the oil is neither as useful nor as significant.

Groundnuts (also known as peanuts or monkey nuts) are another ancient source of oil. These leguminous annual plants were grown in South America from at least 3000–2000BC. The Aztecs were using the nuts as vegetables as well as for making pastes and oils. Groundnut cultivation spread throughout Africa and Asia from the Americas. India, China and the United States of America are now the main producers.

Sesame oil is also ancient in cultivation and use. Unlike the fruits of the aged olive tree, sesame seeds are harvested from an annual plant. Cultivated by the Egyptian Pharaohs, taken from Africa to India, China and Japan, sesame oil was as vital in Asia as olive oil was in the Mediterranean. Sesame oil features in mythology and biblical contexts as a food, for medication and as a beauty product in the same way as olive oil.

Linseed from the flax plant was also known by ancient civilizations. It was used as a food by the Greeks, and Egyptian mummies were bound in linen from the plant. Linseed (or flax) oil was used for culinary purposes, but also as a drying oil in oil painting.

Above: Sesame seeds were first cultivated by the ancient Egyptians.

Rape and soya beans were grown by ancient Chinese civilizations. Rape was also grown in India and soya in Japan. Palm oil comes from trees native to West Africa, producing two types of oil, one from the outer husk of the nut and another from the kernels.

Sunflower seeds have also been used for thousands of years as a source of food. There is evidence to support the cultivation of sunflowers by North American Indians since 900BC. The oil was also used as a hair and skin preparation, and in war paint. In some tribes, Indian mythology included reference to sunflower plants being jealous women who had been turned into plants.

The argan tree is native to Morocco, and oil has been produced from its nuts for thousands of years. Wild Moroccan goats climb the argan trees to feed on the nuts. Traditionally, the Berbers (the indigenous Moroccan people) would collect the undigested stones (pits) of the argan, from the goats' waste underneath the trees. They would grind and press the stones and make a nutty oil that was used in both cooking and for cosmetic purposes.

Above: This oil press was used by the ancient Phoenicians in Tipasa, Algeria.

Above: A hungry goat climbs an argan tree in Morocco to feed on nuts.

OIL THROUGH THE AGES

With worldwide trade, different vegetable oils spread all over the world – including olive oil from Europe, palm oil from West Africa and coconut oil from South-east Asia.

Olive oil is now produced in many different countries, but its use worldwide is relatively small. Production of other oils, namely palm oil and soya oil, is more significant, as they are consumed in much greater quantities.

The spread of olive oil

Olive oil production grew across Europe and beyond. From the 16th century, as the European conquistadors subdued and colonized the New World, olive saplings accompanied missionaries and immigrants on their voyages. Olive cultivation has gradually spread wherever the climate has proved suitable: to California, Chile and Argentina, China, Japan, New Zealand, southern Africa and Australia.

Although olives now grow in many different countries, the Mediterranean still produces most of the world's oil.

Mediterraneans also use the most oil, and consumption is highest of all in Greece, where the average person gets through over 20 litres (35 pints) of olive oil a year.

In recent years, the demand for olive oil has exploded in countries that used to disregard it. In the USA, for instance, consumption rose 2000 per cent between 1980–5, and it continues to increase by about 10 per cent each year. Today there are more than 800 million olive trees, but still about 90 per cent of them grow in their ancient home around the Mediterranean, and many olives are still grown, harvested and processed using techniques that the ancient Greeks and Romans would have recognized.

The production of palm oil

Traditionally produced in West African countries, palm oil was made on large plantations by state-owned slaves.

Above: Oils have been used throughout history for cooking, beauty and health.

Production spread to South-east Asia and Latin America from the 16th century onwards. In Africa, the oil was considered to be so precious that in 1856 in the Kingdom of Dehomey (now the Republic of Benin), a law was passed forbidding the cutting down of oil palms.

During the British Industrial Revolution (in the late 18th and early 19th centuries), palm oil brought back to England by traders was used as an industrial lubricant. It was also used in soap products, such as the Lever Brothers' Sunlight Soap. By the late 1800s, palm oil was the primary export of some West African countries.

Malaysia and Indonesia now produce around 85 per cent of the world's palm oil. Although it is high in saturated fat, which we now know is bad for us, palm oil remains relatively inexpensive, and has a long shelf life,

Above: The olive was, and still is, central to the Italian diet. In the 15th century, Leonardo da Vinci produced sketches for this new and improved olive press.

Above: This painting shows a typical West African palm oil plantation, c. 1845.

and is the most produced and consumed edible oil worldwide. It is used in many developing countries.

Development of other oils

Coconut oil, which is pressed from the flesh of the coconut, was developed as a commercial product by merchants in South Asia in the 19th century. It is very high in saturated fat and not widely used in Western cooking. The Phillipines is the world's largest exporter of coconut oil today.

Spanish explorers probably introduced sunflowers to Europe but it was the Ukraine and Russia that developed sunflower oil to its commercial potential in the early 19th century. In the Ukraine, Alexierka is known as the sunflower capital for its sunflower milling.

Corn oil is a relative newcomer in the oil market. In the early 20th century it was extracted from seeds before they were used for their starch or from the residue in vats after the corn had been fermented for alcohol,

then used for lighting, as a lubricant and in soap production. It was several decades later before it was used as a cooking oil.

The first cottonseed oil was bottled in the USA in the 1880s. A byproduct of the cotton industry, it was used for cooking, but in the 1940s, during World War II, supply could not meet demand, and soya oil was used instead. For several decades, soya oil was the leading vegetable oil in terms of worldwide consumption and production, until palm oil took over. Together, soya oil and palm oil now account for over half of all vegetable oil consumed.

Rapeseed oil had been used in Asia and Europe for centuries, but in the 1970s, Canadian scientists began producing oil from a cultivar of the rapeseed plant. Worried about the negative implication of the name 'rape', they named their oil canola, from the phrase 'Canadian oil, low acid'. Canada is now a major exporter of canola oil.

OIL IN SOAPS

Soaps are produced by a chemical process known as saponification. When a fatty acid reacts with a strong base (alkaline) it produces glycerol and a substance known as the 'salt' of the fatty acid. In relatively simple terms, sodium hydroxide combines with fat to produce a sodium soap – a solid soap. When potassium hydroxide is used a liquid soap is produced. The commercial process includes ingredients for colour, scent and different qualities or uses, such as anti-bacterial properties.

Various vegetable oils are traditionally used in soap making. Olive oil was used for Castile varieties of soap, a term for everyday good-quality 'family' or household soaps. Fancy or toilet soaps were made from palm oil, coconut oil, castor oil or almond oil (lard or tallow, rendered animal fat, especially from cattle and sheep, were also used).

WORLD VEGETABLE OIL CONSUMPTION

Palm oil	31%
Soya oil	29%
Rapeseed (canola) oil	15%
Sunflower oil	8%
Groundnut (peanut) oil	4%
Cottonseed oil	4%
Palm kernel oil	4%
Coconut oil	3%
Olive oil	2%

Source: United States Department of Agriculture (www.usda.gov) 2006

SACRED OIL

From the founding of the city of Athens, to Egyptian goddesses and biblical offerings, oil features in the mythical and religious traditions of many societies.

Oil has been a part of human civilization for thousands of years, and consequently has played a part in many religious rites and rituals.

Oil and ancient mythology

Olive trees were worshipped as sacred in ancient Greece. The tree was said to be a gift to the city of Athens from the goddess Athene. The sacred 'City Olive' tree was planted on the Acropolis.

Olives were traditionally picked and reduced to oil by virgins or pure men. In ancient Greek epic poetry, Homer referred to olive oil as 'liquid gold'. It was used by the gods themselves. In the *Iliad* the goddess Hera uses olive oil as part of a beautifying regime. Aphrodite anoints the body of the Trojan prince Hector with rose-scented olive oil.

In Egyptian legend, Isis was said to have taught man how to cultivate olives. The olive trees were so important that branches were included among the objects placed in the tombs of pharaohs.

In the ancient world, anointing with oil was a general sign of respect for both the living and the dead; guests would be offered oil for their feet on arrival and on leaving.

Oil and religious ritual

The Bible is a very good source of information on the uses of oil, both in society and in a religious context.

Oil was a sign of wealth and olive trees were carefully tended and protected to preserve good fortune. Oil was a currency, used for payment, traded, to meet tithe demands and to

Above: The olive tree, and its oil, feature in ancient Greek mythology.

pay tribute. Throughout the Old and New Testaments, there are references to oil as a food (in bread) and it was also burnt in lamps. Oil was used in ointments for treating the sick, for soothing and healing, as well as for beautification. Honoured guests were anointed with oils on entering a home or before a meal.

The dead were also anointed with oil, a tradition that dated from ancient civilizations who dressed the faces of their deceased with oils to symbolize light and purity, and help them on their way through the underworld.

In religious rituals, oil was, and still is, used in ceremonies to anoint priests during ordination and kings who were enthroned to power. Religious vessels and objects were anointed with oil when they were consecrated. God provided oil for his people and it was withdrawn as a punishment,

Above: Athene's sacred olive tree was planted at the Acropolis in Athens.

Above: Oil has played a part in Christian ceremonies for hundreds of years.

then restored again later. Oil was associated with peace, prosperity, blessings and joy.

Early in the Bible, the significance of the olive tree is established when the dove that Noah sent out from the ark returned bearing a fresh olive leaf as a sign that the waters had subsided and that life could resume. The dove and the olive leaf were a sign of peace.

The Jewish festival of light, Hannukah, celebrates the re-dedication of the temple in Jerusalem. Over the eight days of the festival, an eight-branched candlestick known as a menorah is lit every day to symbolize the oil-burning lamp in the temple that burnt for eight days on sufficient oil for just one day.

The Judaeo-Christian practice of anointing monarchs and priests with oil has its roots in ancient Greek and Roman traditions. The words 'Christ' and 'Messiah' mean 'anointed'.

In Islamic teaching the olive tree is referred to as blessed, neither Eastern nor Western, but providing an oil that glows or shines even without being lit.

Above: The menorah symbolizes the oil-burning lamp that burnt for eight days.

Above: Oil lamps are placed in front of a statue of Hindu goddess Lakshmi.

Olive oil was recommended for anointing as well as eating, and it was reputed to cure disease.

Sesame oil, sometimes known as til oil, is considered scared in the Hindu religion. Lamps containing sesame oil are placed in front of statues of the Hindu gods and goddesses as a form of worship.

Diwali, or the Festival of Lights, is celebrated by Hindus, Sikhs, Buddhists and Jains. During the festival, small lamps called 'diyas' are lit. The diyas consist of a wick inserted into a small clay pot filled with oil, usually coconut oil. The lighting of the diyas represents the triumph of good over evil, knowledge over ignorance and light over dark.

Sacred monuments have traditionally been honoured with oil: libations of oil have been poured over the sacred Omphalos stone at Delphi, the Ka'aba in Mecca and the Jewish Ark of the Covenant. Christian churches are consecrated using holy oil.

HOLY OIL

Olive oil has ancient links with religious rituals. It is used as part of the rituals of anointing but is also burnt in small lamps, to symbolize the process of nourishing the eternal light, providing spiritual nourishment. Just as oil was used to anoint the body of someone who died, candles or lamps lit for the deceased provided a light to lead the way to God.

Above: Libations of oil are used in worship and to consecrate sacred buildings.

FATS: THE FUNDAMENTALS

The right mix of fats, in balanced amounts, promote good health. Limiting the amount of fat we eat is a modern preoccupation, but some fat is an essential part of a balanced diet.

Oils are types of fat. For a healthy balance, we need to know not only how much fat we are eating, but what kind. A little information about what makes liquid and solid fats different, and how they break down in the body, offers insight into how best to use these ingredients.

Fats and oils

Lipid is the general term used for fats and oils. In everyday use, fat usually refers to a solid substance and oil to a liquid. Making a note of the term 'lipid' here is helpful for explaining how fatty foods behave in the body when they are digested, used and eventually passed out of the body.

Chemistry of fats

Fats are triglycerides, made up of fatty acids and glycerol (an alcohol molecule). When digested, lipids are

Above: Vegetable oils are liquid at room temperature, whereas animal fats such as lard or butter are solid.

broken down into fatty acids, their building blocks or components. Fatty acids are organic acids that occur naturally in fats and (like other organic substances) they are made up of carbon, hydrogen and oxygen atoms in different proportions. They consist of a chain of carbon and hydrogen atoms with an acid group at one end. Fats are defined by the chemical bonds between their carbon groups.

Saturated and unsaturated fats

Depending on their chemical make-up, the fatty acids can loosely be grouped into saturated or unsaturated. Saturated fatty acids have all the hydrogen atoms they can hold attached to the carbon atoms in the hydrocarbon chain.

Unsaturated fats have some double bonds between the carbon atoms, which means that if these bonds split up into single bonds, another hydrogen atom could attach to take up the spare bond. Unsaturated fats are found in vegetable oils. The unsaturated fatty acids may be polyunsaturated or monounsaturated.

Saturated fatty acids have the maximum number of hydrogen atoms attached, with two to each carbon atom. These have single bonds between the atoms.

Monounsaturated fatty acids do not hold the maximum number of hydrogen atoms but have one double bond.

Polyunsaturated fatty acids hold fewer hydrogen atoms than monounsaturated fatty acids and have more double bonds.

Polyunsaturated fatty acids

The precise arrangement of the atoms can vary in polyunsaturated acids. The usual, natural arrangement of the atoms is termed cis. There is another arrangement of atoms known as trans, familiar from the term trans fatty acids. The trans arrangement does occur naturally in some foods (to some extent in ruminant fats and milk, that is from cud-chewing species, such as sheep and cows) but this is rare and the more common examples of trans arrangement are found in 'unnatural' products, for example produced when polyunsaturated fat is processed by hydrogenation (as in hydrogenated vegetable oil).

Different levels of saturation

Fat in foods is made up of different combinations of fatty acids. Some fats are more saturated than others. Typically, animal fats are thought of as saturated, while vegetable fats are unsaturated, with some being polyunsaturated and others monounsaturated. Foods can be said to be 'high in saturated fats' or 'rich in monounsaturated fats', indicating the balance of saturated to unsaturated fatty acids in the structure as a whole.

Solid or liquid fats

Saturated fats tend to be solid at room temperature while unsaturated fats are liquid. Butter, lard and meat dripping set at room temperature and become firm when chilled. They have to be heated to melt. By comparison, unsaturated vegetable oils are liquid at room temperature. Place olive oil in a very cold place in the refrigerator and it will thicken but it will not set. Some animal fats are less saturated than others: for example, poultry fat is softer than meat dripping; pork fat is less solid than lamb or beef fat. There are a few exceptions to the animal-vegetable rule: for example, coconut oil and palm kernel oil have high levels of saturated fat. Unsaturated fats that are usually liquid can be processed to make them more saturated and less liquid. Techniques for processing oils so that they are set at room temperature are used to make margarine and spreads. One of the processes is hydrogenation.

Hydrogenation and trans fatty acids

The process of hydrogenation involves pumping hydrogen atoms through the fat at pressure. The available double bonds may then take on some of the hydrogen atoms to become more saturated. Instead of the natural cis arrangement, they may form a trans arrangement, and this results in production of trans fatty acids.

The problem with trans fatty acids is that there is evidence to support links between a high dietary intake and disease; in other words, they may have adverse effects on health, for example, they have been associated with cardiovascular disease and high levels of blood cholesterol.

Hydrogenation was introduced to the manufacture of culinary fats early in the 20th century when margarine was in the process of development. The process is referred to as hardening and it is used for vegetable oils and fish oils in some mass-produced, processed products. As well as margarine, hydrogenated oils are used as cooking fats and are listed among the ingredients on a wide range of products, from spreads, sauces and dressings; biscuits and cookies; pies, pastries and baked goods to hot drink powder mixes.

Above: It is important to include some fats in your diet to achieve the right balance.

Above: Monounsaturated fatty acids can be found in olive oil.

Above: Sunflower oil contains high levels of polyunsaturated fatty acids.

Food manufacturers and retailers are aware of the possible negative effects of a diet high in trans fatty acids. As a result, the widespread use of hydrogenated oils in some cases has, and in many cases still is, being addressed. While some retailers have publicized the fact that they have banned the use of hydrogenated oils and fats in their own-brand products, this is not universal. Check food labels for hydrogenated oils and the ingredients list may come as a surprise. Avoid them where possible and use alternatives.

Essential fatty acids

The body can manufacture the majority of the fatty acids it needs with the exception of two that have to be included in the diet. These are the polyunsaturated alpha-linolenic acid and linoleic acid. Three other important polyunsaturated fatty acids include eicosapentaenoic acid (EPA) and docosahexaenoic acid (DHA) that can be manufactured from alpha-linolenic acid, and arachidonic acid that can be formed from linoleic acid.

Above: Walnuts and their oil are rich in essential omega-3 fatty acids.

Omega-3 and omega-6 fatty acids

The terms omega-3 and omega-6 refer to the type of chemical structure and the position of the double bonds. Alpha-linolenic acid, EPA and DHA are part of the omega-3 group of fatty acids and linoleic acid is one of the omega-6 fatty acids. Omega-3 fatty acids are found in fish oils and some vegetable oils, including walnut, flax seed (linseed) and rapeseed (canola) oils. Omega-6 fatty acids are found in nuts and seeds, and their oils.

Lipoproteins and cholesterol

Fats and fatty acids are not water soluble. In the body, the blood is the transport system for nutrients that have been digested from food. So that digested fats can be carried around in the blood and taken to the cells where they are needed, they have protein molecules (produced in the liver and intestine) attached to them. This produces lipoproteins – the fats (lipids) with protein attached.

There are different types of lipoproteins designed for different functions but their fundamental role is to act as carriers, taking the lipids to the cells where they are needed. There are chylomicrons, high-density lipoproteins (HDL), low-density lipoproteins (LDL) and very low-density lipoproteins.

Potentially 'negative' fats

The LDLs carry about three-quarters of the cholesterol in the blood, taking it to the cells where it is required. When there is an excess of LDLs (more than needed for use in the cells), the cholesterol carried is deposited on the smooth muscle wall of the blood vessels. The LDLs are often referred to as bad fats.

This can lead to a build-up of plaque on the insides of the arteries, which slowly reduces the space for blood flow, which in turn increases the pressure. It is bad news for the body, leading to

Above: Check the labels carefully when buying oils – avoid hydrogenated oils.

increased blood pressure, strain on the heart and cardiovascular disease. Eventually, the vessels may block up completely or bits of plaque may break off and cause a blockage.

Potentially 'positive' fats

HDLs are the carriers that work in the opposite direction. They are responsible for removal of excess cholesterol from the blood, taking it back to the liver where it can be processed and removed from the system. Including a good level of HDLs in the diet helps to control blood cholesterol; they are often referred to as 'good' fats.

Sources of cholesterol

Some cholesterol is obtained from food but the majority is produced by the liver. Even if the diet is not high in cholesterol, the body may manufacture it. There may be some natural disposition to produce more cholesterol – something genetic that means the level of cholesterol is higher than desirable. Having a diet rich in LDLs and low in HDLs will contribute to raised blood cholesterol, but some people who have relatively low-fat diets and are not overweight can still have raised blood cholesterol levels. Eating good fats, maintaining a healthy weight and exercise can help.

Antioxidants

Fat soluble vitamin E is a particularly useful antioxidant, so eating good fats in the right proportion helps to provide antioxidants as well as carriers for taking away excess cholesterol. Fruit and vegetables are also rich in antioxidants, including vitamin C and betacarotene.

Rancidity

The term for fats and oils that have gone off due to oxidation or hydrolysis, and have an unpleasant flavour and odour is 'rancid'. Some fats are more prone to rancidity than others.

Oxidation takes place when fats are exposed to air and the process speeds up in warm, light conditions. Unsaturated fats are more prone to oxidation than saturated fats. This is also one of the reactions that influences the freezer life of foods and some fats become rancid during freezing relatively quickly, especially when salted and/or poorly wrapped with inadequate packing. Bacon is a good example as it will become rancid quickly when frozen.

Oils that are left in half-filled, unsealed bottles will become rancid quickly, especially when left in sunlight. Oil and vinegar bottles should not be left for table use over days or weeks. The salad oil will soon deteriorate and become rancid.

Hydrolysis is a reaction promoted or speeded up by air-borne bacteria, typically in butter that is left uncovered at room temperature. Butter is made up of a certain proportion of water as well as fat. Unlike oxidation, hydrolysis is a reaction between water and fat, with both being broken down. Ghee or clarified butter is heated gently until all the water evaporates, then the salt content and other solids are strained off, leaving a pure fat that does not become rancid easily. Clarified butter is as stable as some oils and more so than others.

Above: Oil that is left out on the table in an unsealed bottle will soon become rancid and unusable. Instead, bottles should be sealed and stored in a cool, dark place.

OILS AND FREE RADICALS

One of the associations between free radicals and fats is in heating. When fats are heated to very high temperatures they begin to break down and form free radicals. Sometimes, it is not one-off heating to a high temperature that causes the problem, but repeated heating, for example when deep-frying in the same fat several times over. Gradually the fat breaks down and develops off flavours.

Just as some fats are solid (butter) and others are liquid (oil), some can be heated to a higher temperature than others before they begin to break down, develop a nasty smell, bitter flavour and burn. This is called the smoking point, when the fat begins to give off a smoke and smell acrid. Some oils break down at lower temperatures than others, making them less suitable for cooking. Not only does the flavour of the oil change but the nutritional value is reduced and free radicals are formed. Overheating oils is not recommended and some oils are best reserved for cold preparation.

The smoking point is lowered each time the oil is heated or used for cooking, as some breakdown occurs at moderate temperatures. Frying in the same oil over and over also leads to breakdown, bad flavours and free radical production. If oil is used more than once, it should be strained to remove bits of leftover food that will overheat and burn with repeated frying.

Vegetable oils reach smoking point at different temperatures. Oils which are unrefined often have a lower smoking point than refined oils. This makes them unsuitable for cooking methods requiring high temperatures, such as deep-frying.

OIL PRODUCTION

The Romans invented a screw press to extract oil from olives many centuries ago. Since then, technologies have advanced, but the basic principles of oil production have not altered.

When oil is made, the basic ingredient — seeds, nuts or fruit — from which the oils are taken undergoes a series of different stages of production. The precise method is important for the quality of the oil. The ingredients are cleaned, milled and pressed to extract the oil, which is then allowed to settle, and is filtered or processed to remove unwanted water and debris. Technology has taken over for mass production but some artisan products are still produced following old-fashioned methods.

Traditional oil extraction

The first stages of oil production involve separating the seeds or fruit from the unwanted debris, such as stems, twigs and leaves, and then washing the basic product if

Above: This 15th-century engraving shows the production methods that were commonly used to make Italian olive oil in the 1600s. The olives are crushed by a stone press.

Above: Wooden presses were originally used to extract the oil from olives.

necessary, for example when processing olives or avocados. Seeds and nuts have to be hulled or shelled.

Traditional methods of production for olives involved crushing the ingredients to a paste by hand and mixing the paste to warm it and help release the oil. The paste was then spread on mats, stacked and pressed to extract the oil and all the juices. When the paste was allowed to stand, the oil would float on top of the water-soluble juices. It would then be separated.

Modern oil production

The basic principles used in oil production today remain the same but the technology has advanced.

Crushing can be done by steel rollers or tooth grinders rather than using stones. Malaxation, the process of mixing, is controlled to allow a certain level of natural heating (27–8°C) but not too much. The higher the temperature and the longer the mixing time, the greater the yield of oil will be. The liquid oil mixture is then separated from the solids, known as the pomace. Instead of standing it in tanks to allow the oil to surface and separate, a centrifuge is used to speed up the process. Centrifugation involves spinning at high speed so that the oil is separated from the solids and other juices. Centrifugation may be used instead of pressing in some processes.

Depending on the type of equipment and techniques used, the oil may be filtered under carefully controlled conditions to remove bits of pulp and shell, sometimes referred to as foots. Further processing may be necessary for some oils, including, for example, refining, bleaching and deodorizing.

Cold or hot pressing

For cold pressing of oil the basic ingredients are not heated before they are ground. The process does not extract the maximum amount of oil and the remaining oil, in the form of a 'cake', can be further processed. Cold pressed oils are not refined but they are filtered. The yield is far lower but premium oil is produced, with fuller flavour and greater nutrient retention. Some cold pressed oils have a sediment that may be retained for its flavour. The residue oil in the cake from the first cold pressing can then be extracted by hot pressing, resulting in inferior oil.

For hot pressing, the ingredients may be heated before crushing or alternatively the crushed paste may itself be heated, for example by steaming or roasting. This helps the cells to break down and release oil but the heating also releases other constituents of the basic ingredients. The oils have to be further processed to remove unwanted substances that have been released by heating.

Some processes involve adding solvents for maximum extraction, resulting in a product from which the oil has to be separated from the solvent mix by further heating.

Traditional classification of oils

Oils were traditionally classified by several criteria, such as origins, properties and uses. The categories of animal, vegetable and mineral oils covers fish oils, the familiar edible vegetable oils and types of fuel oil respectively. Another means of

Above: The types of olive oil can be recognized by colour; cold pressed olive oil has a distinctive green shade.

grouping oils is by their chemical properties, either fixed or volatile. While fixed oils do not rapidly evaporate (or do not evaporate at all), volatile oils evaporate comparatively quickly and these include the essential oils now familiar for aromatherapy.

Oils can be referred to as 'drying' or 'non-drying', terms that will be familiar to artists who use oil-based painting media. Drying oils are also a key component of many varnishes. When drying oils are exposed to air, they form a tough surface. A good example is boiled linseed oil, which is excellent as a wood preservative; this is inedible due to the metallic dryers that are used in its production. Boiled linseed oil is thicker than the normal oil and has a shorter drying time. Ordinary linseed oil does not have the same properties and will not dry as quickly – it will remain sticky when applied in quantities that cannot be soaked up. Other commonly used drying oils include tung oil, poppy seed oil and walnut oil.

Above: Bottles are filled with oil on a production line in a modern oil factory.

DIRECTORY OF OILS

There is an amazing variety of oils suitable for all sorts of applications, from frying food to beautifying the complexion, and from the well-known Mediterranean olive oil, to Moroccan argan oil and South-east Asian coconut oil. Many familiar, relatively inexpensive oils can work their magic beyond the kitchen, using methods that have been passed down through many generations. This section provides a guide to oils that are readily available for household use – from cooking and cleaning, to relaxing and reviving.

Left: Vegetable oils vary in colour, from very pale, almost colourless yellows to rich browns.

A WIDE RANGE OF OILS

Many different oils are now available to the modern consumer. It is important to know how to choose the right oils, as well as to store them correctly and eventually dispose of them safely.

Many types of edible oils are used all over the world. Historically, different countries have used oils pressed from whatever raw material was available to them. Olive oil has been used in the Mediterranean for centuries, while palm oil is popular in Africa, Asia and South America.

Some oils are used specifically for their flavour. Sesame oil and groundnut (peanut) oil are used in Asian cooking, for example in stir-fried dishes, to impart a nutty flavour. Other oils, such as grape seed oil, are favoured for their very mild flavour.

Other edible oils which are available include coconut (or copra) oil, soya oil, sunflower oil, rice bran oil, safflower oil, corn (or maize) oil, flax (or linseed) oil, wheatgerm oil, avocado oil, pumpkin seed oil, hemp seed oil and poppy seed oil. Some oils are more common than others, although most can be sourced from specialist and online retailers.

Vegetable oil

Cooking oils which are sold simply with the label 'vegetable oil' are usually a blend of a variety of oils such as sunflower, corn or soya oil. The exact contents of the oil will be listed on the bottle.

Not all vegetable oils are edible. Examples of such oils include processed linseed oil and tung oil. These oils can be put to use around the house, outside of the kitchen.

Buying oils

The majority of larger supermarkets sell a good selection of oils, from basic vegetable cooking oil, through various seed and nut oils to an increasingly broad variety of olive oils. Larger supermarkets usually have a good stock turnover but some of the expensive less-common oils may have been stored for rather longer than ideal in a warm light environment. Healthfood shops and delicatessens usually offer a different selection, often including the more unusual oils; buy from busy shops to ensure that the stock is not old.

Storing oil

Oil becomes rancid when stored in a warm, light place for any length of time. Dark, airtight bottles are the best containers. Buy oil in quantities that will be used fairly quickly, especially the more expensive types, such as nut oils, that are unlikely to be used as extensively. Make space in a cool cupboard, for example, well

TASTING OILS

By assembling a number of distinctive oils you can try a comparative tasting of oils in the same way as you would taste wine. It may be interesting to try several varieties of the same oil, for example olive oil, of which there are many types available to buy.

Pour each oil into a small glass. Warm the oil in your hands and swirl it around the glass, then inhale to test the aroma. Take a small sip and note the flavours, which for olive oil can range from fruity, mellow or grassy to peppery and pleasantly bitter. The oil should not taste fatty in the mouth, or acidic in the throat. Cleanse your palate between oils by eating a slice of apple.

Above: Taste small amounts of different oils and compare flavours and aromas.

Above: Warm oil in your hands before tasting it, to release the flavour.

away from cooking appliances, washing machine or tumble dryer, refrigerator or freezer (they give out quite a lot of heat). A cupboard on an external wall is usually cooler.

Oils can be stored in the refrigerator if there is no cool cupboard in the kitchen but as space can be a problem, limit this for best olive oils or other oils that may not be used on an everyday basis.

It is pointless having a wide variety of 'treat' oils that are used infrequently, only to discover that they have gone off when the bottles are half empty. It is better to decide on one or two good oils for frequent use and have phases for one or two 'special' oils, buying small bottles to use fairly quickly.

Check the storage instructions and use-by dates on individual bottles because oils vary in their processing, which in turn influences their keeping quality.

Special care should be taken when storing oils that have been flavoured with herbs or other ingredients. Flavoured oil should always be strained; if remnants of the herb remain in the oil, harmful bacteria can develop.

Disposal of oil

Oil that has been used for cooking, for example after deep-frying, should not be poured down the sink as it can cause blockages in pipes. Instead, it should be allowed to cool, then transferred to a suitable container, sealed and disposed of with other household waste. This is not very environmentally friendly, however, and there are alternatives. Recycling schemes for cooking oil exist in some areas, and small amounts of vegetable oil may be safely added to a compost heap.

Above: Sesame oil has a distinctive flavour and is commonly used in Eastern cooking.

Above: Olive oil is widely used in many traditional Italian dishes.

Above (clockwise from top): Pumpkin seed oil, hazelnut oil, grape seed oil and corn oil.

OLIVE OIL

Centuries old, olive oil is widely used for culinary and other products, from the best dining to candles and the finest soaps and beauty products.

Used regularly in the ancient world, olive oil, along with bread and wine, is one of the oldest foods in existence. Away from the Mediterranean climate, animal fats were historically favoured as cooking fats until relatively recently. Its popularity has now grown worldwide, and today olive oil is the oil for which there is the most variety and choice available.

Much has been written about the history, extraction, types and uses of olive oil. Tree, olives and oil feature throughout the history of the ancient world, and olive oil continues to be a source of research and development. The global food market has evolved such that oil aficionados may sample the products not only of different countries but of specific producers and from the fruit of identified varieties. In the past, simply using virgin olive oil was indicative of culinary sophistication. Nowadays the average supermarket offers a choice of at least a few different types of olive oil while larger outlets display shelves of olive oils from many different countries. The following background information provides a starting point for exploring the nuances of the many olive oils that are now available.

Places and production

The Mediterranean countries are still the focus for olive cultivation. Spain, Italy and Greece are key producers of olive oil. Turkey, Syria, Tunisia, Morocco, Portugal, France and California contribute, as do Australia and South Africa to a far lesser extent. Producing countries export their olives and oil but they also import olive oil; for example, Italy is the second largest producer in the world, but the Italians themselves consume more oil than they produce – 11 litres (19 pints) per person each year – and Italian law allows the producers to import oil for blending and bottling without stating its original source. So, while some of the finest single estate oils come from Italy, most of the blended oil labelled 'bottled in Italy' on the supermarket shelf is actually Spanish, Greek or Tunisian in origin. Consumption per person is significantly higher in Greece than in either Spain or Italy.

Olive varieties

There are many varieties of *Olea Europaea*, the cultivated olive tree, with national and regional names. Greece is known for its Koroneiko and Kalamata olives. From Spain, Picual, Hojiblanca and Verdial olives are popular in Andalusia;

Above: Olive oil has been part of the Mediterranean diet for thousands of years.

Above: With its renowned health-giving qualities, olive oil is now used worldwide.

Above: There are many varieties of olive, each producing different types of oil.

Above: Olives that have been hand-picked will be used to make the finest and most expensive olive oils.

Above: Olives in higher branches are picked using a small rake. The olives bruise easily so great care must be taken.

Above: Large nets are placed on the ground around the trees in order to catch the olives as they fall.

further north, Cornicabra olives are the focus of Castilla La Mancha; while in the Aragon and La Rioja regions, the Arbequina and Empeltre olives dominate.

Italy is well known for its regional oils with their separate characteristics resulting from the different soil types and weather patterns. Tuscany is the major area for olive oil production and one of the major exporters to other regions. Italy provides oils with all types of characteristics, from the light and fruity to robust and spicy. Moraiolo, Frantoio, Leccino and Pendolino are among the main varieties of olive grown.

Quality and flavour

The flavour of the olive oil depends on a number of contributing factors. The variety of olives and the conditions under which they are grown obviously influence the characteristics of the oil. Just as in wine production, the type of soil and weather also play their part. The skills of the producer in planting and caring for the trees as well as in judging the harvest and the ripeness of the olives are also extremely important for the quality of the oil.

Traditional planting allows about 32 trees to the acre. Left to grow, the trees can reach a height of 10–15m (30–50 feet), living for up to a thousand years. Careful pruning and feeding are necessary to prevent the tree from growing to full height while preserving the fruit-bearing branches that yield a good crop for practical harvesting. Pest control involves preventing insect and fungal attacks that are typical of all tree nurturing, but the olive moth is the particular pest that will consume the leaves and buds.

Harvesting and processing

The olives have to be ripe but not too ripe to produce good oil, and the way in which they are handled influences the finished product. The harvesting season varies according to the region and conditions but typically takes place between November and February.

Harvesting methods and production techniques should be as gentle as possible for producing the finest oils. Protecting the olives against damage is important to help prevent the flesh from deteriorating in quality between picking and processing. Hand-picking is still practised for the finest oils. Producers who go to these lengths will use the fact that the oil is extracted from hand-picked olives as a major selling point. Other methods include netting the ground under the trees to catch the fruit as the trees are shaken or when the olives drop naturally, as hitting the ground causes bruising.

The olives are loaded into baskets and taken for cleaning, sorting and extraction. The time lapse influences the quality of the oil and alters the acidity of the olives, which naturally begin to ferment once they are picked. While a short period of storage may improve the end product, too long will spoil it and allow off flavours to develop. The olives may be taken to a central processing plant or the grower may also have the necessary processing equipment.

Grades of olive oil

Olive oil is graded according to its acidity levels as well as the quality of the flavour and aroma. Extra virgin and virgin olive oils are from the first pressing, are produced by cold pressing and do not contain any refined olive oil.

Extra virgin olive oil has a maximum acidity of 0.8% (according to IOOC standards) or 1% by some regulations. The oil must have a good flavour and odour.

Virgin olive oil must have good flavour and odour, with maximum acidity of 2%.

Ordinary virgin olive oil has a maximum acidity of 3% but the flavour is generally not as good as the superior types.

Refined olive oil is a term that may be used for oil obtained from virgin live oils by refining methods that do not alter the structure of the oil. Refined olive oil has a very low acidity level, for example as low as 0.2%.

Pure olive oil or olive oil is oil that contains a blend of refined and virgin olive oils. They generally have a maximum acidity of 1.5%. These oils do not have the full flavour or range of flavours of the extra virgin or virgin oils.

Lampante is the term for oil with an acidity level higher than 3.3%, poor flavour and unpleasant aroma. This type of oil is not suitable for consumption but can be used for non-food use. It can be refined, bleached and deodorized to produce an edible oil that can be blended with better quality oils. Alternatively, the refined oil may be used in food processing or for packing (for example, when olive oil is used in the canning industry).

Manufacturer's terms

Apart from the formal grading, other terms and names are used on labels.

Light olive oil or mild olive oil are terms for refined olive oils, meaning that they are bland in flavour. Light olive oil may be used in blended oil products, for example, where olive and sunflower oils are mixed, and this will be stated on the label. 'Light' or 'lite' does not refer to the fat content of the oil.

Blended olive oil indicates that the oil is made up of oils from different varieties of olive and different regions. Blending is used to maintain a consistent standard. When the term applies to a mixture of oils, the other types of oil (sunflower or rapeseed oil, for example) must be listed.

Early or late harvest oils refer to the ripeness of the olives from which the oil has been extracted. The flavours will be quite different if fully ripe olives are used compared to green olives.

Unfiltered oil will retain some of the paste, which is thought to give more flavour, and produce a sediment.

Cold pressed oil

In regulation terms, cold pressed refers to the maximum temperature of 27°C/80°F that the olive paste may reach during the mixing process.

Above: Kalamata olives after the first pressing with traditional stone rollers, still used in oil presses in Greece.

Above: Modern production lines allow oil to be produced on a much greater scale to meet the growing demand.

Above: Strict rules govern the grading of olive oil, set by the International Olive Oil Council in Madrid.

Above: Olive oil was used as a moisturizer by the ancient Romans, and still makes a quick and easy treatment for dry skin.

Above: Traditional olive oil soap, or 'savon de Marseille', has been manufactured in France for hundreds of years.

Above: Inexpensive olive oil can be used around the house as a natural cleaner, restorer and furniture polish.

The paste warms naturally during mixing, increasing with the length of mixing; applying a maximum temperature controls the quality of the oil. Increasing the temperature gives a higher yield but the flavour is inferior.

Flavour characteristics

The flavours of different olive oils vary significantly and many of the terms used to describe the analysis of the oils resemble those used in wine tasting.

Oils may be 'robust', 'fruity' and 'full flavoured', especially when extracted from olives that are just about ripe. Full fruity flavours are positive. Lighter flavours may include hints of 'green' or slightly 'grassy' tones and these may be positive, combined with lighter fruit flavours, such as apple. 'Sweet' is used when the oil does not have sharp or bitter qualities to its flavour. This term does not mean sweet in sugary terms but by comparison to sharp. The words 'spicy' or 'peppery' are used to describe complex flavours reminiscent of sweet spices or of slightly hotter pepper.

Sometimes olive oils are criticized for poorly balanced flavour or for unwelcome characteristics that may include bitterness or earthy tones. Cucumber-like flavours are not good and may indicate that the oil has been stored for too long (but has not been exposed to air or high heat, so has not gone rancid). Some oils may be described as 'harsh', 'astringent' or 'rough'.

USE

- In all aspects of cookery: as a cooking medium, for stir-frying, as an ingredient in baking, for marinades, sauces and dressings, in dips with other ingredients.
- To make your own herb- or spice-flavoured oils.
- As a simple dip on its own to moisten good bread.
- As a beauty product. Olive oil is a traditional hair and skin conditioner.
- As a health supplement.
- To restore, clean and polish woods, metals or leather.
- As a lubricant for stiff locks and hinges.

SETTING STANDARDS

The International Olive Oil Council (IOOC) based in Madrid, Spain, is responsible for the standards that are applied, terms, descriptions and the labelling of different oils.

Founded in 1956, following the first International Olive Oil Agreement, the original aims of the organization were to modernize and promote the growing of olives and olive products and to draw up international procedures for the sector. The United States has its own set of rules for olive oil producers but there is some pressure for producers there to follow the IOOC guidelines.

NUT OILS

Strongly flavoured, rich in vitamin E and with high levels of monounsaturated fat, versatile nut oils are a great addition to any store cupboard.

Almond oil

Along with olive trees, almond trees have been cultivated since ancient times in the Mediterranean basin, but with the emphasis on the nuts themselves rather than for oil extraction. Almonds are still grown across the Mediterranean. California is one of the world's major growing areas, as are Italy and Spain. Almonds contain 40 to 60 per cent oil.

There are two types of almond oil extracted from different botanical species. Sweet almond oil has a mild taste, while bitter almond oil has the distinctive, strong almond flavour. **Sweet almond oil** is light and flavourless. Although it can be used as a cooking medium for food, it is quite expensive. It is usually

Above: Almonds have been cultivated for thousands of years. Oil extraction only forms a small part of their use.

used in baking, as a final dressing for hot foods, as a base for salad dressings, or in sweet dishes where a distinct flavour is not required. It is useful for greasing moulds for setting cold dishes, especially sweet foods. Away from the kitchen, sweet almond oil is widely used as a base oil for massage and aromatherapy uses. Being light and odourless, it makes a good carrier for the aromatic essential oils.

USE
- In sweet baking mixtures, such as American muffins and cookies.
- To drizzle over food, such as salads or charcuterie, to moisten it and introduce a hint of oil without additional flavour.
- As a base for light salad dressings.
- As a carrier oil in aromatherapy.

Bitter almond oil In nature, bitter almonds contain a substance known as amygdalin that can produce the toxin prussic acid or hydrogen cyanide. In processing the bitter almonds, the prussic acid is eliminated, making the oil safe for culinary use. Bitter almond oil is used in very small quantities as a flavouring in sweet cooking, including baking and confectionery.

USE
- As a flavouring ingredient in sweet baking, blancmange or desserts.

Almond oil with flavouring In addition to the two oils, some almond oils are based on sweet almond oil with

Above: Sweet almond oil is expensive and is not widely used in cooking. It is often used in aromatherapy as a carrier oil

flavouring added, to produce an oil with a light almond flavour. This can be used to make dressings with a light almond flavour or drizzled over a variety of savoury or sweet foods.

Groundnut (peanut) oil

Peanuts, groundnuts or monkey nuts are popular for a wide variety of culinary uses across Asia, Africa and South America. They have long been used in cooking and as a source of oil.

Groundnut oils may be cold pressed, in which case they retain the distinct flavour of the nut, or refined to produce a light oil that is virtually flavourless. The refined oil is ideal for deep-frying as it has a high smoking point, therefore it can be heated to a high temperature. The cold pressed oil

Above: Groundnut oil is often used in Asian cooking, such as stir-fries.

Above: The sweet flavour of hazelnut oil makes it ideal for use in desserts.

Above: Walnut oil is rich in vitamin E and is perfect for use in salad dressings.

can also be used as a cooking medium in dishes that benefit from its flavour, but it cannot be heated to the same high temperature so is not suitable for deep-frying. Both types of oil can be used for roasting, baking, shallow frying, stir-frying, or in sauces and dressings, including mayonnaise.

Butterine oil was an old-fashioned term for the oil obtained from a second cold pressing of groundnut oil, and this was used as an edible oil as well as a fuel. (Butterine was also a name applied to margarine.) Hot pressed oil from a further processing was used in the soap-making industry. Groundnut oil was often used to adulterate olive oil and lard. In turn, it was adulterated by the addition of cotton seed oil, poppy seed oil and rapeseed (canola) oil.

USE

- Refined or cold pressed groundnut oil for speedy stir-frying (the former is preferable when cooking at very high temperatures).
- As part of a coating or to brush food before baking or grilling (broiling).

- In salsas, dressings, dips and sauces, such as satay dipping sauce, for a distinct peanut flavour.
- Refined oil for deep-frying to give crisp, dry results.
- As a base oil for massage.

Hazelnut oil

This has the distinct flavour of the nuts and is excellent for drizzling over a wide variety of hot or cold foods as a dressing. It can be used in either savoury or sweet dishes. Hazelnut oils are produced in Turkey, Greece, Italy and Spain as well as in the United States. It is valued primarily for its nutty flavour.

USE

- To drizzle the oil over leafy salads, cheese, meat and poultry as a dressing.
- On its own or mixed with a lighter oil in salad dressings.
- In light dips and to finish creamy sauces for flavour.
- In sweet creams and with syrups or honey to make dessert sauces.
- As a carrier oil for massage.

Walnut oil

Growing wild across Asia and the Balkan countries, walnut trees were probably first cultivated in Turkey and Iran. The trees are now cultivated in many countries all over the world, from Russia, East Asia and across the Mediterranean countries to France, where there is a long history of cultivation and oil production. California is now the major producer.

Walnuts contain about 70 per cent fat, with a high percentage of monounsaturated fatty acids. They are also a good source of vitamin E. Walnut oil is expensive and used for its flavour rather than as a cooking medium.

USE

- To drizzle over foods, savoury or sweet, in dressings and sauces for hot or cold dishes.
- In marinades and dressing for cooked meats, cheese or fruit.
- In savoury dips or sweet creams, syrups or sauces.
- To flavour sweet dishes, moisten cakes or pastries.
- As a base oil for massage.

Pistachio nut oil

The pistachio tree is native to the Middle East, Asia and India where it has grown wild for centuries. It was probably first cultivated in Asia and introduced to Europe by the Romans, then on to America. The nuts contain up to about 55 per cent oil, of which 50 per cent is monounsaturated. The nuts are valued as an ingredient in savoury cooking, with oil extraction being relatively unimportant. The oil is well flavoured and expensive, and most suitable for use as a dressing or for final flavouring in savoury and sweet cooking.

USE

- To drizzle over cooked foods, cheese or salads.
- In dressings, sauces and dips.
- To flavour sweet dishes, creams and syrups.

Macadamia nut oil

Also known as Queensland nuts, macadamia nuts originated in Australia, where they are a traditional

Above: Pistachio nut oil varies in colour, depending on production methods, from a distinctive green to a paler yellow.

Aboriginal food. They were introduced to Hawaii in the late 19th century and they are also cultivated in South America and Africa. The nuts are widely valued for their flavour and texture. They are rich in vitamin E and monounsaturated oil and contain about 70 per cent fat.

Above: Pistachio nuts are generally valued more for use in cooking than for the expensive oil extracted from them.

These expensive nuts yield an equally expensive oil (only a small percentage of the nuts are designated for oil extraction) that is light coloured and textured, with the distinct yet delicate flavour of macadamia nuts. Although light, there is a certain rich, creaminess to the flavour.

Macadamia nuts are used in commercial bakery and confectionery products and the oil is valued in the cosmetics industry. It is reputed to have excellent moisturizing and antioxidant qualities that are valuable in products for skincare.

USE

- As a dressing, drizzled over cheeses and cooked foods.
- In salad dressings.
- To flavour and enrich sauces and dips.
- With chocolate and sweet ingredients for making rich sauces and dressings for desserts, ice creams and cakes.
- In natural skincare remedies.

Above: Macadamia nuts are processed ready for oil extraction.

Above: Macadamia nuts are expensive, and as a result so is their oil.

OTHER EDIBLE VEGETABLE OILS

There is now a huge variety of edible vegetable oils available to us – from the less familiar African palm oil and coconut oil to store-cupboard staples such as sunflower and corn oil.

Palm oil

A native of Africa, cultivation of the oil palm has spread to Asia and South America, and it is one of the richest sources of oil on the planet. There are two distinct types of palm oil and it is important to differentiate between them. *Palm oil* is the term for oil that is extracted from the fibrous pulp of the fruit, which contains about 50 per cent fat. This oil contains about 50 per cent saturated fatty acids and about 40 per cent monounsaturated fatty acids with about 10 per cent polyunsaturated fatty acids. The oil is strongly coloured red or orange due to its high betacarotene content. The oil also provides vitamin E. The betacarotene content is destroyed when the oil is refined, which it is for the majority of use but some unrefined palm oil is available. Unrefined palm oil can be

Above: Oil is pressed from coconut flesh (copra) which is dried in the sun.

Above: Coconut oil is predominantly used in South Asian cooking.

Above: The high betacarotene content in palm oil gives it a distinct red colour.

found in some African stores and blended with other oils, for example with rapeseed (canola) oil. The palm oil may be referred to as 'red palm oil' or 'red palm fruit oil'. *Palm kernel oil* is extracted from the kernel of the nut and it is quite different in composition from palm oil, with a high percentage of saturated fat (about 80 per cent). Unlike the oil from the fruit, the kernel oil is pale in colour and does not contain betacarotene.

Palm oil and palm kernel oil are both widely used in margarine and commercial food production, including fats, vegetable oils and shortenings, peanut butter, baked goods, ice creams and snacks.

USE

- As a cooking medium for frying, roasting or baking (check the ingredients if using a blended oil).
- In marinades, dressings or sauces.

Coconut or copra oil

Copra is the term for dried coconut flesh, which contains 65 to 70 per cent oil. Coconut oil is extracted from the dried flesh and widely used in commercially prepared food products, such as baked goods and margarine. It is also used as a cooking oil, especially in some regions of India.

Coconut oil contains high levels of saturated fatty acids. The oil is also used in the cosmetics industry and in soap production. It can also be used as a fuel for diesel engines.

USE

- As a cooking medium, especially when frying.
- To add an authentic flavour to South Asian dishes such as sauces or curries.
- In natural skincare remedies.
- As a carrier oil for massage.

Above: Soya beans are native to East Asia. They are harvested and used in tofu and soy sauce as well as oil.

Soya oil

The soya beans that are used to make soya oil are better known in the domestic sense as whole beans, or for the many products that use the whole bean, such as tofu and soy sauce. Although soya beans contain a relatively modest proportion of oil (14 to 24 per cent) it is highly unsaturated and it is one of the most important vegetable oils. Soya oil is widely used in blended vegetable oils, margarines and in food products.

USE
- As a cooking medium, for frying, roasting and baking.
- For marinating.
- As a base for salad dressings or in sauces.

CASTOR OIL

Extracted from the seeds of the castor oil plant, castor oil is not an 'edible' oil in the culinary sense but an old-fashioned purgative or laxative.

Above: Soya beans, and their oil, are a good source of protein and often used as a dairy substitute.

Sesame oil

An ancient oil-producing crop, the cultivation of sesame originated in Africa, from where it spread to India and throughout Asia, on to America. Sesame seeds are the fruit of an annual plant and they are just as important whole as they are for oil production.

Above: Plain sesame oil is light in colour but has a strong sesame flavour.

They contain about 50 per cent fat that is mainly polyunsaturated. The oil has the great advantage of a long shelf life as it does not turn rancid easily.

Plain sesame oil is light in colour and mildly but very distinctly flavoured. Toasted sesame oil is the product of the toasted seeds and this is far darker in colour and much stronger in flavour.

Sesame oil is added to margarine and used in cooking, usually in a relatively modest proportion to bring flavour, while a plainer oil that does not burn as easily is used for the purpose of frying. Both plain and toasted sesame oils are often used in Asian dishes.

USE
- In frying or stir-frying to prevent the oil from overcooking at high temperature and becoming bitter. Add a little to cooking part-way through.
- In marinades and sauces.
- In small quantities in salad dressings and cold sauces or dips.
- As a base oil for massage.

Above: Oil made from toasted seeds is darker in colour than plain sesame oil.

Sunflower oil

Native Americans have used sunflowers for thousands of years in their breads and as a source of oil. The seeds were introduced to Europe by early explorers and on to Russia in the 19th century. The Ukraine takes the credit for invention of the first sunflower mill specifically for oil extraction. Russia and France are important producers of sunflower oil outside the United States.

Sunflower seeds vary in their composition but the fat content is between about 25 to 50 per cent, with a high proportion of polyunsaturated fatty acids. Before olive oil increased in popularity with the increased interest in monounsaturated fat, sunflower oil was promoted as the healthy alternative to saturated fats.

Sunflower oil is widely used in spreads, shortenings and margarines as well as in mixed vegetable oils.

USE

- As a cooking medium for frying, roasting, baking and in baking mixtures.
- In salad dressings and sauces.

Above: Sunflower oil has a light flavour, and is used most commonly as a frying oil.

Above: Fields of yellow rape can be seen throughout Europe, as well as in Canada, the United States, China and India.

Rapeseed (canola) oil

The rape plant is also known as cole, coleseed or colza as well as canola. The name 'canola' originally derived from the phrase 'Canadian oil, low acid'.

The two main varieties of rape that are widely cultivated include oilseed rape or swede rape (*Brassica napus*)

Above: Sunflower seeds are pressed for oil, which has a high vitamin E content.

Above: The mild oil produced from rape is yellow in colour, and is a good source of essential fatty acids.

and turnip rape, toria or sarson (*Brassica rapa*). While there is evidence of rape having been grown across China and India some 2000 years ago, it is relatively new on the commercial oil scene. The oil was originally used in Asia and Europe in lamps. Rape seeds contain about 40 per cent oil, which is highly unsaturated with a large proportion of monounsaturated fats. The oil is rich in both omega-6 and omega-3 fatty acids.

The bright yellow rape crop is now grown across Canada, China, India and Europe. The resulting oil is yellow in colour, mild and light but with a distinct flavour. Its aroma is mild and almost floral, with some similarity to a sweet corn kernel aroma. The oil is used as a cooking oil and in margarine production. It is used widely in Scandinavia.

USE

- As a cooking medium, for frying, roasting or baking.
- In dressings and marinades.

Rice bran oil

A pale yellow oil with a light texture and bland flavour, rice bran oil is the oil extracted from the germ and husk of rice. It is highly unsaturated (about 75 per cent) with a high proportion of monounsaturates and a good vitamin E content.

The filtered cold pressed oil has a high smoking point of 490°F/254°C and therefore it is recommended for frying and as a cooking medium. It has been used in vegetable ghee, which is sold as an alternative to dairy ghee.

The light flavour of rice bran oil makes it suitable for salad dressings and sauces, especially when highly flavoured nut oils are added or if the flavour accent of other ingredients is required. It is a popular cooking ingredient in Japan and China.

USE

- For high-temperature cooking methods such as stir-frying and deep-frying.
- In dressings and sauces where a mild flavour is preferred.

Above: Corn or maize oil retains a distinct corn flavour when cold pressed.

Above: Safflower oil is similar to sunflower oil in appearance and nutritional content.

Safflower oil

A relative newcomer to modern global cooking oils, safflowers were originally grown for their brightly coloured petals rather than their seeds. The petals were sold as 'false saffron', as a cheaper alternative to the real thing, and also made into dye. The safflower plant resembles a thistle or small globe artichoke with bright yellow, orange or red coloured petals. Safflower seeds produce an oil with a high level of polyunsaturated fats and a light flavour.

USE

- In gentle cooking and in cold dishes, dressings and sauces.
- As a base oil for massage.

Corn or maize oil

As well as oil from the germ, which contains about 25 to 50 per cent fat that is largely polyunsaturated, corn provides starch. Cold pressed, unrefined corn oil has a distinct flavour of corn, while the refined oil is far lighter in flavour and colour. Both types are used as a general cooking

Above: Safflowers are grown commercially in more than 60 countries worldwide.

medium in the same way as sunflower or vegetable oil as well as for salad dressings, salsas and dips.

USE

- In marinades, batters or baking for distinct flavour (cold pressed corn oil is best for this).
- To contribute a distinct corn flavour to fried foods, such as chicken or corn fritters.
- For all general cooking when a strong flavour is not required (use refined corn oil).
- For salad dressings, salsas, dips and other cold dishes (both types can be used for this).

Wheatgerm oil

Cold pressed wheatgerm oil is valued for its high vitamin E content. It is largely polyunsaturated with a light mild flavour.

The oil is expensive and generally promoted as a supplement rather than a culinary ingredient, with manufacturer's recommendations provided for daily dosage. Its high vitamin E content also makes it useful for beauty products, such as hand and face creams and face masks or moisturizing scrubs.

USE

- In cold dishes, such as salad dressings or sauces.
- In natural skincare remedies.
- As a base oil for massage.

Cottonseed oil

With a content of about 50 per cent polyunsaturated fatty acids, cottonseed oil is used as a general cooking oil and in the production of margarine. It is often used in manufacturing potato crisps (US potato chips) and other snack foods. Early in the 20th century, several grades of cottonseed oil were produced, some with a 'pleasantly nutty taste' according to an early guide to groceries. Cottonseed oil was widely used to adulterate olive oil.

USE

- As a cooking medium, for frying, roasting or baking.

Above: Wheatgerm oil is used mainly as a health supplement and in cosmetics.

Above: Oil has been extracted from the seeds of the flax plant for centuries.

Flax or linseed oil

Linseed oil is extracted from the seeds of the flax plant. It has a long history of cultivation and culinary use in Russia and Eastern European countries, especially Silesia, a region of Poland, and the Czech Republic and Slovak Republic. Linseeds are small, shiny, brown seeds that contribute an excellent texture to roasted seed mixtures, breads and similar baked goods. The oil is not widely used for cooking (better known in the paint industry and for wood preservation) but it is now popular for the polyunsaturated fats it contains, especially omega-3 and omega-6.

The cold pressed oil has a distinctive aroma, golden colour and bitter flavour. It becomes rancid relatively quickly and manufacturers recommend that opened bottles should be stored in the refrigerator. It is not suitable as a cooking medium but can be used as a dressing on hot or cold foods. Traditionally, it is used to dress potatoes and other vegetables.

Healthfood brands, including those sold through supermarkets, suggest taking the oil as a supplement rather than using

Above: Linseed oil is not used in cooking, and is better as a dressing ingredient.

it in cooking. Its distinctive bitter flavour makes this oil a difficult ingredient that needs to be carefully balanced and it is something of an acquired taste.

Russia, India and South America were the traditional producers of linseed oil a century ago. Baltic and Black Sea oils were obtained from Russia, East India oil from India and River Plate oil from South America. The Baltic oil was the better quality of the Russian and Indian oils as it was less likely to be adulterated by foreign seeds of other plant varieties. The oils were imported into Britain for use in the painting and decorating trades.

USE

- The oil may be taken in small doses as a health supplement.
- As a dressing – mix with apple juice and honey, adding a little lemon for a refreshing sharpness that is different from the bitterness of the oil.
- To drizzle over mild yogurt and soft cheeses; add honey or syrup for a little sweetness to make a creamy dressing for salads or vegetables.

Avocado oil

The flesh of an avocado contains about 30 per cent oil, with a good percentage of monounsaturated fats. Avocado oil has a light yet distinct avocado flavour and the unmistakable richness of the avocado. It makes an excellent dressing for salads that include avocado and it can be used in guacamole or other avocado dips. Refined avocado oil is also produced and recommended for use as a cooking medium.

Avocado oil can be used for home-made beauty products, in moisturizing face masks, hand and foot treatments.

USE

- In cold dressings, especially with light vinegars, lemon or lime juice.
- To drizzle over salami, cold meats, smoked fish or cheese.
- As a dressing for fruit, such as papaya, in savoury salads and dishes.
- For moistening grilled (broiled) fish.
- For enriching avocado dips or other dips and sauces.
- In mayonnaise for a richer flavour.

Above: Grape seed oil varies in flavour and colour, from strongly flavoured green to much lighter refined varieties.

Above: Avocado oil can be added to dips such as guacamole for a rich flavour.

Grape seed oil

While cold pressed grape seed oil is green with a pronounced flavour, there are also very light and flavourless, refined grape seed oils. Refined grape seed oil can be used as a cooking medium; however, it is an expensive choice. Cold pressed oil can also be used for cooking but it is better suited to cold preparation.

A good source of vitamin E and alpha-lineolenic acid, grape seed oil is also used commercially in cosmetics and can be useful for making moisturizing treatments.

USE

- To moisten fish and poultry before baking in the oven.
- As a medium for frying.
- In marinades for fish, poultry, meat or vegetables.
- In salad dressings, dips and salsas.

Plum seed oil

A French product, this is extracted from the kernels of plums grown for drying to produce prunes d'Agen. The oil is highly unsaturated with about 70 per cent monounsaturated

Above: Pumpkin seed oil retains the mild nutty flavour of the seeds.

and 20 per cent polyunsaturated fats. The key characteristic of the oil is its fabulous almond-like flavour. The expensive oil is used as a dressing or flavouring ingredient.

USE

- In dressings, dips and sauces, savoury or sweet.
- For drizzling over fruit or desserts, combining with syrups or stirring into creams.

Pumpkin seed oil

The cultivation of pumpkins originates from South America and Mexico, with certain species being cultivated for their seeds. The seeds contain about 45 per cent fat, largely polyunsaturated. The oil is dark green and viscose with a full flavour that is mild and nutty. The oil is expensive and useful for cold preparation.

USE

- As a dipping oil, with bread.
- In salad dressings, sauces and dips.
- To drizzle over cooked foods or dishes, or into soups.

Argan oil

The argan tree (*Argania spinosa*) is an evergreen that is native to Morocco. The oil is extracted from the seeds of its fruit. The cold pressed oil is rich in vitamin E and alpha-linoleic acid. It has a distinct flavour and nutty, toasted aroma, with a hint of sharpness but not bitterness.

Argan oil is expensive and useful for cold cooking or dressing hot or warm food rather than as a cooking medium.

USE
- As a dipping oil for breads.
- In salad dressings or to drizzle over cooked meats, vegetables, cheese or cooked dishes, such as tagines.

Hemp seed oil

A member of the cannabis family, hemp is the name for a plant from which yarns are obtained. Cannabis is illegal in most countries, and seeds used for oil have to be free from the substance associated with cannabis used as a drug. The golden-coloured oil available from supermarket ranges is refined and lighter than the deep green cold pressed oil with a strong flavour. The golden oil has a distinctly green or grassy flavour, with a slight bitterness but less so than linseed oil.

Hemp seed oil can be used in warm or cold dishes but not as a cooking medium. The oil is a good source of omega-3 and omega-6 fatty acids.

Above: Oil is extracted from the seed of the hemp plant, commonly used for fabrics.

Above: Poppy seeds are edible and have been used in cooking for many years.

USE
- As a supplement. The oil may be taken in small doses following the manufacturer's instructions.
- Combined with sweet and sharp ingredients to make dressings.
- Drizzled into portions of vegetable soups or dips that will benefit from the 'freshness' of its green flavour.
- Complemented with lime or lemon rind and juice.

Poppy seed oil

The seeds of the poppy flower have been used in cooking for thousands of years. Poppy seed oil has a subtle flavour and is used in dressings and as a condiment for bread rather than as a cooking oil.

The oil is rich in omega-6 fatty acids, and is used in the production of skincare products and soaps, as well as in the manufacture of paints and varnishes.

Above: Roasted argan seeds are ground into oil by hand. The oil is expensive, and has a distinct nutty flavour.

Above: The evergreen argan tree is native to Morocco. Each fruit contains 1–3 of the small oil-rich seeds.

USE
- As a condiment or dip for bread.
- In salad dressings.

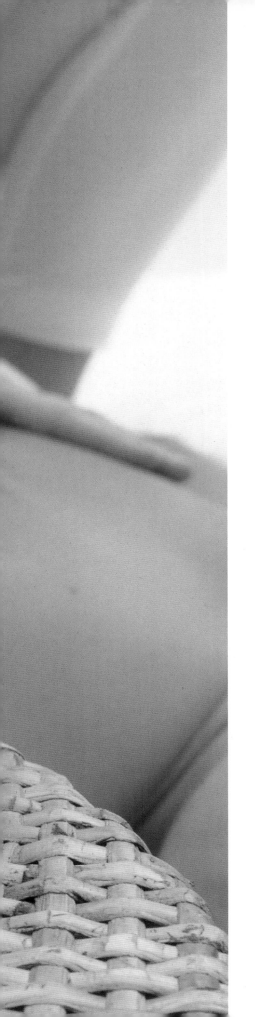

OILS FOR HEALTH AND HEALING

Oils play numerous roles in promoting good health, featuring in the ancient practice of massage therapy as well as playing an essential part in a balanced diet. Adding just a few drops of a favourite essential oil for massage will transform a simple vegetable carrier oil, giving it relaxing, soothing or healing powers. Oils are vital for our health and wellbeing, and can be used to treat various ailments such as head lice, insomnia, colds, headaches, spots, burns, bites, bruises or stings.

Left: Massage oils consist of a small amount of a chosen essential oil which is blended with a vegetable carrier oil.

VEGETABLE OILS AND MASSAGE

One of the main contemporary uses of oil as an aid to health and wellbeing is as a medium in massage. Massage aids relaxation, helps tired or strained muscles and can ease health problems.

Massage is one of the oldest therapies, and oils have been used in massage for thousands of years. Massage reduces tension and stress, can ease tired muscles and aching limbs, energize your whole being and even alleviate certain complaints.

Carrier oil basics

In massage, small amounts of different essential oils are dissolved in a vegetable carrier oil to make a blend that makes it easier to move the hands continuously on the skin without dragging or slipping. Choosing an appropriate carrier oil will heighten the dynamic nature of a massage and can have specific benefits, such as helping to guard against heart disease or inflammatory diseases such as arthritis. It can also help to boost the immune system. When used in massage, vegetable carrier oils can be absorbed into the blood stream through the skin's pores.

Vegetable oils are made up of essential fatty acids and contain the fat-soluble vitamins A, D, and E. Some oils also contain large amounts of gamma linoleic acid (GLA), useful for the treatment of PMS. The fatty acid compounds help to reduce blood cholesterol levels and strengthen cell membranes, slowing down the formation of fine lines and wrinkles and helping the body to resist attack from free radicals.

Heat-treated oils lose their nutritional value, so always use a cold pressed, unrefined vegetable oil

Above: Almond oil (front) is gentle and is suitable for most skin types, including very dry or dehydrated skin. Grape seed oil (left) has almost no smell, so is ideal for blending with essential oils. Olive oil (back right) is easy to get hold of, but it does have a strong smell that can be difficult to disguise.

as a carrier for essential oils. Likewise, use a certified organic vegetable oil, as this guarantees that no chemical fertilizers, pesticides, or fungicides have been used in its production. The darker the colour and stronger the odour, the less refined the oil, so it will be richer in health-giving properties.

Vegetable carrier oils

The following oils can be used alone, or as a carrier for essential oils. Once opened, store in the refrigerator.

Almond oil A good source of vitamin D. It is suitable for all skin types, but is especially good for dry or irritated skin.

Sweet almond oil is one of the most versatile carrier oils. It is easily absorbed and is a warming, light oil. It can help to relieve muscular pain and stiffness.

Avocado oil Easily absorbed into the skin, it is excellent for dry or mature skin. It can help to relieve the dryness and itching of psoriasis and eczema. It blends well with other oils, and its fruity smell may influence which essential oils you choose.

Coconut oil Light coconut oil is used extensively in traditional Indian head massage. It is easy to use and blends well with essential oils.

Grape seed oil A non-greasy oil that suits all skin types. This oil is not usable as a cold pressed oil but is widely available in a refined state and is best enriched with almond oil.

Hazelnut oil Its astringent qualities make it a useful carrier oil for oily and combination skins.

Olive oil Too sticky for massage, but excellent in a blend for mature or dry skin. Use the best quality virgin, extra virgin or cold-pressed oils as these contain high levels of unsaturated fatty acids that are nourishing for dry skin and hair.

Groundnut (peanut) oil This is best when unrefined, but rarely available. Its refined form makes a good base oil for massage, but is best enriched with a more nutritious oil if you require more than just a slippage medium.

Safflower oil This is light and penetrates the skin well. Cheap and readily available in an unrefined state, it is a useful base oil.

Sesame oil When made from untoasted seeds, sesame oil is very good for treating skin conditions. It has sunscreening properties and is used in many suncare products. Use commercial preparations with a stated SPF number. In Ayurveda, sesame oil is very popular for massaging the head and body. It helps to strengthen, condition and moisturize the skin and hair. It is a balancing oil and can help to reduce pain and swelling.

Sunflower oil A light oil rich in vitamins and minerals. It can be enriched by the addition of more exotic oils.

Walnut oil This contains small amounts of essential fatty acids and has a pleasant, nutty aroma.

Wheatgerm oil Rich in vitamin E and useful for dry and mature skin. It is well known for its ability to heal scar tissue, reduce stretch marks, and soothe burns. It is too sticky as a massage oil, so add small amounts of it to a lighter oil. It should not be used on people with wheat intolerance.

Above: Generally, add 3–5 drops of essential oil to 10ml/2 tsp carrier oil. Check labels for individual instructions.

Above: Warm a small amount of your blended oil in the palms of your hands to test the fragrance.

BLENDING OILS FOR MASSAGE

Experiment with different types of carrier oil to achieve the ideal blend for your massage style. Try adding a teaspoonful of another carrier oil as well as the essential oils for a highly personal mixture. It is worth remembering that even the weather affects the state of our skin, and in the winter central heating and cold temperatures will cause it to dry out. These variations can be accommodated by changing the exotic carrier oils used to enrich each blend.

Rub a little of the blend between the palms of your hands to warm it, then test the fragrance before beginning the massage. It may require slight adjustment before you are happy with the result.

Before you begin blending the oils, wash and dry your hands and make sure that you have all the bowls and bottles you need, and that all your utensils are clean and dry. Have your essential oils at the ready, but leave the lids on the bottles until they are required.

Measure out approximately 10ml/2 tsp of your chosen carrier oil and gently pour it into your blending bowl.

Bearing in mind the correct ratio of essential oil to carrier oil (generally 10ml/2 tsp base oil to 3–5 drops of essential oil, but check labels for guidance) add the first essential oil a drop at a time. Add remaining oils a drop at a time and mix gently with a clean, dry cocktail stick or toothpick, to blend.

ESSENTIAL OILS

An essential oil is the essence of a plant, the plant's life force distilled for use. The oils are extracted from many different parts of a plant – leaves, flowers, fruits or other material.

Essential oils should not be applied undiluted to the skin, but should be mixed first, in the right proportions, in a vegetable carrier oil. Massage is just one use for essential oils, but when there is not time for a massage, adding a few drops of oil to a bath or foot soak is also beneficial. Both soaking and inhaling make use of the different health-giving properties of the oils.

Useful essential oils for health

Benzoin The tree gum is available in dissolved form. It is the key ingredient in Friar's Balsam. Known for clearing the head and useful for healing, for example to treat dry rough skin or chilblains.

Bergamot This is obtained from the rind of a citrus fruit. It is a bright and uplifting oil, which is helpful for overcoming anxiety and depression. It is also used to assist in the treatment of urinary tract infections such as cystitis.

Eucalyptus This is good for muscular pain and is effective against coughs and colds, both as a preventative and as a remedy.

Frankincense Known for its sedative and anti-inflammatory qualities, this calming oil also has antiseptic qualities and it is helpful for easing coughs and bronchitis. Frankincense is thought to help slow and deepen breathing, which makes it helpful for overcoming anxiety.

Geranium The aroma of geranium oil makes it valuable in beauty potions and lotions. It is antiseptic with astringent characteristics and known as a balancing oil that can be helpful for skins that may be too oily or dry. It has antidepressant qualities and is said to be calming.

Juniper Known for its antiseptic, astringent, cleansing and diuretic qualities, juniper is thought of as a detoxifying oil for the body and useful for settling and sorting racing minds. Mentally, it can act in a restorative sense. It is also thought to help reduce cellulite.

Lavender One of the most popular oils, lavender is calming, soothing and balancing. It is also antiseptic and healing. Lavender is versatile and one of the most useful oils for helping with all types of healing, including improving bad skin and soothing minor burns. Lavender is also helpful for inducing sleep or overcoming insomnia. A lavender bath, massage or just a little massaged into the wrists before bed helps with sleep. It is also a good oil to use for massaging tired muscles and for a soothing, de-stressing shoulder massage or head massage.

Lemon A lively oil that is thought to help stimulate the immune system, lemon is also anti-bacterial and good for cleansing cuts. It is also astringent and antiseptic. It is an uplifting and stimulating oil, helpful for overcoming depression.

Marjoram A warming oil that can be used to clear the head and ease coughs associated with a cold. Marjoram is a sedative and sleep-inducing, and it can numb the senses. It is also thought to aid digestion and ease indigestion. It is said to be helpful for easing menstrual cramps.

Myrrh Anti-inflammatory and expectorant, myrrh will help to ease bronchitis, catarrh, coughs and colds. Also good for digestive problems, infections of the mouth and throat, and skin conditions.

Above: Essential oils should be mixed with a vegetable carrier oil before they are applied to the skin. Always check the label for the recommended quantity to use, as it can vary.

Neroli An orange-blossom oil, this is another oil that is helpful for calming and reducing anxiety. It has anti-depressant and soothing qualities, and is helpful for promoting relaxation and sleep. Neroli is useful in skin lotions and it is thought to stimulate skin regeneration and help to reduce the effects of ageing.

Palma rosa This is good for the skin and is also used to relieve stiff and sore muscles, while calming the mind and uplifting and invigorating the spirits. Often used in soaps, perfumes and cosmetics.

ESSENTIAL OIL WARNINGS

• Always use essential oils diluted.

• Never take essential oils internally unless professionally prescribed.

• Do not use the same essential oils for more than one or two weeks at any one time.

• For problem or sensitive skin, dilute the oils further. If any irritation occurs, stop using them.

• Some oils, such as bergamot, make the skin more sensitive to the sun, so use with caution.

• Essential oils should be used only according to medical and professional aromatherapy advice during pregnancy as some should be avoided.

• Similarly, because the oils can be powerful and in some cases the effects can be cumulative, those suffering from medical conditions or who suspect they have a problem should seek medical advice.

• Essential oils can be helpful and can complement some other treatment but using the appropriate oils and techniques is vital.

• If you are unsure about the suitability of an oil, always seek the advice of a qualified aromatherapist.

Patchouli An oil with a distinct aroma, patchouli is known as antiseptic and anti-inflammatory. It can be used for skincare and for haircare, to promote good scalp health and help overcome dandruff. Patchouli is also thought of as an aphrodisiac and anti-depressant.

Peppermint Uplifting and stimulating, peppermint oil is thought to help overcome fatigue and ease headaches. It has antiseptic and antispasmodic properties, and is helpful for overcoming indigestion and upset stomachs.

Pine Antiseptic and decongestant, pine is known as one of the head-clearing oils that are excellent for inhaling or adding to baths to relieve a bad cold. Along with eucalyptus, it is useful for relieving sinus congestion and it is also thought to help sooth coughs. Pine is a stimulating oil, and is helpful for promoting circulation and for treating aching muscles.

Rose One of the favourite essential oils, valued for skincare and beauty products. Rose oil is anti-depressant and calming, soothing and helpful for overcoming insomnia. It is widely used in aromatherapy to treat a wide variety of conditions, from menstrual to sexual problems, depression and nervousness. Rose oil is thought of as a general tonic among essential oils.

Rosemary Rosemary oil is a powerful stimulant. It is known for astringent and head-clearing qualities, which include decongestant properties as well as an ability to promote clear thinking. It is also useful for relieving muscular pain. Rosemary oil is also useful for haircare, in rinses and conditioners. Being a strong stimulant, rosemary oil should not be used by anyone with epilepsy.

Rosewood A calming and soothing oil that has antiseptic properties, rosewood is useful for overcoming stress and

Above: There are many essential oils which are beneficial to the health.

depression. It is anti-bacterial and antiseptic and a mild oil for beauty products and skincare.

Sandalwood A soothing, anti-depressant and aphrodisiac oil, sandalwood has antiseptic properties and has long been used to help overcome urinary tract infections. One of its main uses is as a balancing oil for skincare, overcoming both dryness and excess oiliness.

Tea tree This oil has strong antiseptic properties with a matching aroma. It has excellent antimicrobial and antifungal action. It is a powerful stimulant to the immune system.

Thyme An antiseptic oil that stimulates digestion and helps to clear congestion. Thyme is thought of as a balancing oil, being calming or stimulating, helping to promote sleep, relieve fatigue and overcome depression.

Ylang-ylang An oil with a sweet, heavy aroma, ylang-ylang is anti-depressant, calming and sleep inducing. It is said to be useful for reducing blood pressure, promoting relaxation and easing stress. It is also said to have aphrodisiac qualities.

NATURAL REMEDIES WITH OILS

Vegetable oils can be used to treat a wide variety of minor ailments whether used on their own or mixed with essential oils, herbs or other ingredients.

Aromatherapy compress

Applying an aromatherapy compress will help to ease bruising, pain or arthritic joints. Use a warm compress for general pain, aching or arthritic joints, and a cold compress if the area is inflamed, swollen or hot.

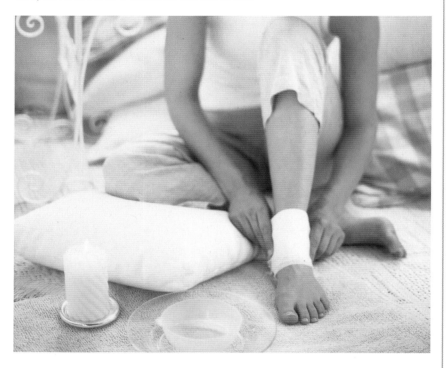

INGREDIENTS

10ml/2 tsp grape seed oil

4 drops geranium essential oil

3 drops bergamot essential oil

3 drops clary sage essential oil

1 Put the grape seed oil in a small bowl, then add the essential oils and blend.

2 Add the blended oils to a bowl of hot or very cold water and mix.

3 Soak a small towel or facecloth in the water, wring it out and hold on the affected area. Replace the cloth often so that the temperature remains constant.

Health tip

To make an aromatic ice-pack, freeze the same essential oil and water mix in an ice-cube tray. Never apply ice directly to the skin. Wrap the aromatic ice cubes in a cloth and apply to the affected area.

Oil pulling

This is an alternative therapy which has its origins in ancient Ayurvedic writings.

A small amount of oil is held in the mouth every morning for up to 20 minutes. Sesame or sunflower are the recommended oils to use. This must be done on an empty stomach, and the oil must not be swallowed.

The oil is swished around the mouth, pulled through the teeth (hence the name of the therapy) and then spat out after a maximum of 20 minutes. The mouth is then rinsed out thoroughly with 2 or 3 glasses of cold water.

The idea behind the therapy is that toxins are pulled out of the body into the oil, then removed from the body altogether when the practitioner spits out the oil.

It has been claimed that oil pulling can help with many illnesses and complaints, from headaches and migraines to eczema and ulcers, although there is currently no scientific evidence to support this.

Energy boosting massage cream

Use this cream for a soothing and revitalizing massage session.

INGREDIENTS

20ml/4 tsp almond oil

40ml/8 tsp avocado oil

20ml/4 tsp rosewater

5ml/1 tsp lecithin granules

10g/^{1}/4fl oz beeswax

8 drops each pettigrain and peppermint essential oils

1 Put the almond oil, avocado oil and beeswax into a ceramic or stainless steel jug (pitcher). Stand in a pan that is half-filled with water.

2 Heat on a low temperature, stirring occasionally until the wax melts. Remove the jug from the water.

3 Add the lecithin and beat the mixture vigorously, then stir in the rosewater.

4 Allow the mixture to cool slightly, then mix in the pettigrain and peppermint essential oils and transfer into a clean screw-top jar. Store in a cool dark place. This will last for about 12 treatments.

Rheumatism liniment

A liniment is a liquid preparation, often made by mixing a herb oil with a tincture. For rheumatic pains, aching joints and tired muscles, rub this liniment gently into the affected areas. It should be applied to the skin at body temperature. Do not apply to broken skin.

INGREDIENTS

6 garlic cloves, crushed

300ml/^{1}/2 pint/1^1/4 cups olive oil

For the juniper tincture

15g/^{1}/2oz dried juniper berries

250ml/8fl oz/1 cup vodka

50ml/^{1}/4 cup water

1 First make the juniper tincture. Add the dried juniper berries to a glass jar. Pour in the vodka and water. Put the lid on and leave in a cool, dark place for 7–10 days (no longer), shaking occasionally.

2 Stain through a sieve (strainer) lined with kitchen paper before pouring into a sterilized glass bottle. Seal with a cork. The tincture will keep for up to 2 years.

3 Put the crushed garlic cloves in a bowl and pour over the olive oil.

4 Cover the bowl with a piece of foil and stand it over a pan of simmering water. Heat gently for 1 hour. Check the water level regularly and top up as necessary.

5 Strain the oil, allow to cool, then stir in the tincture of juniper and pour into a stoppered bottle. This will keep for several months if stored in a cool dark place.

Variation

Juniper tincture can be taken internally but contains alcohol so should not be given to children. Take no more than 5ml/ 1 tsp, 3–4 times a day, diluted in water or fruit juice if preferred.

Insomnia massage spray

This massage spray contains chamomile and lavender essential oils, which are both prized for their relaxing and sedative properties. Giving yourself a foot massage at bedtime will help you to relax and drift off to sleep.

INGREDIENTS

25ml/5 tsp grape seed oil

25ml/5 tsp almond oil

20ml/4 tsp jojoba oil

10ml/2 tsp rosewater

10ml/2 tsp glycerine

20 drops each lavender and chamomile
    essential oils

1 Mix the grape seed oil, almond oil, jojoba oil, rosewater and glycerine together in a small bowl.

2 Stir in the essential oils, mixing well. Transfer to a clean spray bottle.

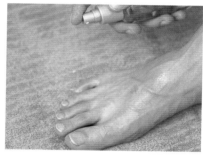

3 Sit in a comfortable chair or lie down in bed. Spray a little of the mixture on to each foot, or on to a large, clean tissue and wipe both feet with the tissue.

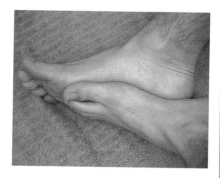

4 Use your feet to massage each other.

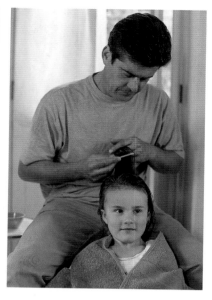

Head lice treatment

A common problem among school-age children, head lice (nits) can be difficult to eradicate. Using essential oils offers a natural solution.

INGREDIENTS

30ml/2 tbsp coconut or almond oil

5 drops lavender essential oil

5 drops geranium essential oil

5 drops eucalyptus essential oil

1 Combine all the ingredients and apply the mixture all over the head and hair, massaging it in well.

2 Cover the head and leave the oils in for a minimum of 4 hours, although overnight is better.

3 To remove the oil, massage shampoo into the hair before applying water, then wash and rinse as usual. Comb through the hair with a lice comb.

4 Repeat the whole process after 24 hours and again after 8 days. This will give you the opportunity to treat any lice that have hatched since the first treatment. The treatment should be stored in a dark glass jar for up to 12 months.

Hair repair

Olive oil has been used as a hair and scalp conditioner since ancient times. Use this hair repair as an intensive, deep-conditioning treatment once a month for hair that is thinning.

INGREDIENTS

30ml/2 tbsp olive oil

10ml/2 tsp wheatgerm oil

8 drops rosemary essential oil

6 drops patchouli oil

1 drop lavender essential oil

1 Mix olive oil and wheatgerm oil together in a small glass bowl.

2 Add the rosemary, patchouli and lavender essential oils.

3 Massage the oil mixture into the hair and leave overnight. (Protect your pillow with a towel.) Rinse off thoroughly the next day.

Health tip

Warm a little olive oil in the palm of your hand and massage into the scalp once a week to treat dandruff.

OLD-FASHIONED OIL REMEDIES

Olive oil was traditionally sold in small corked bottles in chemists' shops and used in small amounts for many treatments that would not be recommended today, for example, a little warm olive oil was trickled into the ear to help relieve earache. Some of the following are included out of fascination rather than recommendation, as today we have alternative remedies.

BOWEL PROBLEMS One of the traditional reasons for taking spoonfuls of olive oil as a supplement was to overcome constipation. In more extreme circumstances castor oil was taken. A dose of castor oil was also recommended for anyone suffering from diarrhoea, the idea being to expel as much as possible of the cause of the problem before taking a remedy to prevent the problem. One cure for diarrhoea, to be taken after the oil and said to 'act like magic' was a mixture of 5ml/1 tsp salt and 15ml/1 tbsp vinegar.

SCALDS A mixture of equal quantities of lime water and linseed oil were used to treat scalds – a piece of linen cloth was soaked in the mixture and applied to the scalded area.

SORE THROAT AND LOST VOICE When a sore throat leads to a lost voice, people used to gargle with warm olive oil. The idea of gargling with pure olive oil seems a little extreme, but using a mixture of half oil and half hot water (from a boiling kettle) would seem soothing. Follow by slowly sucking a teaspoon of honey, which is excellent for the throat as it is a natural antiseptic.

PATCH TESTING

It is a good idea to patch test essential oils before using them. Add 2 drops of essential oil to 2 drops of carrier oil.

Massage into a patch of the delicate skin on the inside of your arm and leave for 6 hours. If no reaction occurs it is probably (but not certainly) safe to use. If a reaction occurs, apply lots of carrier oil to the area to help neutralize the effects. Do not rub the area.

Lavender and eucalyptus vapour rub for colds

A blocked nose is a misery when suffering from a cold and prevents a sound night's sleep. This decongestant rub has a warming and soothing action and should be rubbed gently on to the throat, chest and back at bedtime, so that the vapours can be inhaled throughout the night. It can also be inhaled in boiling water. Known particularly for its head-clearing properties, eucalyptus oil is common in decongestants and inhalers.

INGREDIENTS

50g/2oz petroleum jelly
15ml/1 tbsp dried lavender
6 drops eucalyptus essential oil
4 drops camphor essential oil

1 Melt the petroleum jelly in a bowl over a pan of simmering water. Stir in the lavender and heat for about 30 minutes.

2 Strain the through muslin (cheesecloth), leave to cool slightly, then add the eucalyptus and camphor essential oils.

3 Pour into a clean jar and leave until the rub is set. Store in a cool, dark place and use within 1 month.

Headache remedy

Lavender essential oil can help to soothe headaches.

Add a few drops of lavender oil to a clean tissue and sniff throughout the day to relieve headaches.

Essential oil inhalant

Inhaling steam scented with aromatic essential oils is an excellent way to relieve the congestion of a cold or blocked sinuses. Try the combination of oils below, or use other decongestant essential oils such as cinnamon, eucalyptus, lavender, lemon, marjoram, peppermint or pine.

INGREDIENTS

600ml/1 pint/2 1/2 cups boiling water

5 drops eucalyptus essential oil

2 drops camphor essential oil

1 drop citronella essential oil

1 Pour the boiling water into a large bowl.

2 Add the eucalyptus, camphor and citronella essential oils to the water.

3 Sitting at a table with the bowl in front of you, lean forwards over the bowl, covering your head with a towel to keep the steam in.

4 Inhale the scented steam for about 5–10 minutes, but stop if your face becomes overheated or you feel uncomfortably warm.

Burns, bites, bruises or stings

This simple oil-based mixture can be used to soothe minor burns, bites, bruises or skin blemishes.

Pour 25ml/1 1/2 tbsp sunflower oil or almond oil in a small glass jar. Add 15ml/ 1 tbsp lavender essential oil and mix together thoroughly.

BUYING ESSENTIAL OILS

Always buy essential oils from a reputable supplier to be sure of obtaining the pure and concentrated oils. Good quality essential oils are expensive, so be wary of any exceptionally cheap oils. They should be purchased in small quantities and stored in small dark-glass bottles with airtight tops in a cool, dark place. A dropper and small funnel are useful for measuring and mixing.

Essential oil burner

Plant essential oils have a powerful effect – breathing in their vapours can be relaxing, restorative or uplifting.

Put a few drops of oil on a handkerchief and keep it on your pillow overnight. For a more controlled and concentrated method, an essential oil burner is the answer.

Spot treatment

To combat troublesome spots, use a mixture of lavender oil and olive oil.

Add 1–2 drops of lavender essential oil to a little olive oil and apply to the problem area with a small piece of cotton wool (cotton balls).

NATURAL BEAUTY WITH OILS

There are very good reasons for using vegetable oils and essential oils as part of a regular routine without giving up a favourite face freshener or body lotion. The natural moisturizing and conditioning qualities of oil have been recognized for thousands of years. Occasional oil-based beauty products such as hair conditioners, face packs, foot lotions and hand conditioners can all be made at home cheaply and easily for use as a special treat to boost everyday beauty regimes.

Left: Vegetable oils are very good for the skin and hair – they are rich in vitamin E and are natural moisturizers.

OIL AND BEAUTY

A variety of oils can be used to prepare home-made natural beauty remedies. Use them in relaxing bath oils, moisturizing body creams and face masks.

Oils have long been associated with beauty and skincare. Egyptian men and women oiled their skin and hair throughout the day with almond oil scented with frankincense and myrrh. The ancient Greeks and Romans certainly knew that olive oil was good for the skin, and used it after baths, before meals, before and after exercise and before and after journeys. The Greek philosopher Democritus had a simple recipe for health: honey on the inside and oil on the outside.

Oils have been used for many centuries to ease tired muscles, soften rough skin and soothe abrasions. They are also a traditional conditioning treatment for the hair, giving it added body and shine.

Oil and skincare

Women through the ages have used oils made from local produce in their beauty regimes. Moroccans have used argan oil to keep their skin soft and supple for centuries, while Mediterraneans favour olive oil. Olive oil contains polyphenols, antioxidants that can help to slow down the ageing process, as well as preventing and repairing sun damage to the skin. Sweet almond oil is used in many skincare products; it is a very light oil which is easily absorbed into the skin. All vegetable oils contain vitamin E which is very good for the skin as it can help to reduce the appearance of scars, stretch marks and age spots. Oils can be gently massaged into the skin as they are, or can be used to make rich, moisturizing face packs and body lotions.

WHEATGERM OIL

Wheatgerm oil is rich in vitamin E and considered to be very good for the skin, although shouldn't be used by anyone with a wheat sensitivity.

For an intensive conditioning scalp treatment, warm 15ml/ 1 tbsp each of wheatgerm and olive oil and massage gently into the scalp. Wrap a warm towel around the head and leave for 10 minutes before rinsing off the oil.

Above: Oils can be fragranced with a few drops of a favourite essential oil.

Above: Any vegetable oil can be used for beauty treatments. Olive oil was traditionally used by the ancient Greeks and Romans; the Egyptians used scented almond oil.

Above: Argan oil is used by Moroccan women to keep their skin soft and smooth.

Above: Oil can be used to make nourishing body lotions and creams.

STORING BATH OILS

Home-made bath oils are best stored in coloured glass bottles, as exposure to light can cause the essential oils to deteriorate; plastic bottles should be avoided as the oils can react badly with the plastic.

Oil and cleansing

Bath oils are a wonderful beauty boon for those with dry skins. They float on the top of the water, and coat your entire body with a fine film when you step out of the bath. Ready-made bath oils are available, or use a few drops of any vegetable oil, such as olive, corn or groundnut (peanut) oil. Adding a few drops of a scented essential oil as well will give you a wonderfully fragrant bath.

Since the 6th century, French soap manufacturers have used olive oil to make their famous Marseilles soaps. Oils are also widely used in body scrubs and cleansers.

Oil and haircare

Any vegetable oil is suitable for conditioning the hair. Warm it slightly in the palm of your hand before massaging into the scalp. Wear a plastic shower cap for 20 minutes before shampooing and rinsing off the oil; the heat from your head will help the oil penetrate the hair shaft.

Above: Keep your favourite oils in the bathroom along with other beauty products.

OILS FOR BATHING

When combined with a blend of essential oils, bath oils can influence mood and health. Olive oil soaps can be enriched with nut oils for a luxurious bathtime treat.

Seductive rose and sandalwood bath oil

Certain essential oils have an undeniably sensuous fragrance – this is certainly true of both rose and sandalwood essential oils. When combined with sandalwood oil, rose oil creates a warm, spicy aroma.

INGREDIENTS

100ml/3^{1}/$_{2}$ fl oz almond oil

20ml/4tsp wheatgerm oil

15 drops rose essential oil

10 drops sandalwood essential oil

1 Pour the almond oil and wheatgerm oil into an opaque glass bottle.

2 Add the rose and sandalwood essential oils and gently shake to mix.

3 Run a warm bath and add 15ml/ 1 tbsp of the oil to the water before you step in. Store in a cool, dark place and use within 1 year.

Beauty tip

Rose oil and sandalwood oil are costly to buy but a little will go a long way.

Milk and honey bath oil with rosemary

Milk is well known for its cleansing and lubricating qualities when applied to the skin. The addition of a little shampoo makes this a dispersing oil which does not leave a greasy rim around the bath.

INGREDIENTS

2 eggs

45ml/3 tbsp rosemary herb oil

10ml/2 tsp honey

10ml/2 tsp baby shampoo

15ml/1 tbsp vodka

150ml/^{1}/$_{4}$ pint/^{2}/$_{3}$ cup milk

1 Beat the eggs in a small bowl, then add the rosemary oil and mix.

2 Add the remaining ingredients to the bowl and mix together thoroughly.

3 Pour the bath oil into a clean opaque glass bottle.

4 Add 45ml/3 tbsp to the bath and keep the rest chilled, for use within a few days.

Grapefruit and coriander bath oil

A stimulating and refreshing combination of oils, this acts as a great reviver, especially when you are recovering from a cold or treating tired muscles after an exercise session at the gym.

INGREDIENTS

100ml/3^{1}/$_{2}$ fl oz almond oil

20ml/4tsp wheatgerm oil

30 drops grapefruit essential oil

30 drops coriander (cilantro) essential oil

1 Carefully pour the almond oil and wheatgerm oil into an opaque glass bottle.

2 Add the grapefruit and coriander essential oils and gently shake to mix.

3 Run a warm bath and add 15ml/ 1 tbsp of the oil to the water immediately before you step into the bath.

4 Store the bath oil in a cool, dark place and use within 1 year.

Olive oil and lavender soap

Enrich a block of naturally green Marseilles olive oil soap with nut oils and finely ground almonds and then scent it with lavender essential oil to make pretty guest soaps. Use heart-shaped moulds for a romantic touch.

INGREDIENTS

Makes about 4 soaps

175g/6oz Marseilles olive oil soap

25ml/1fl oz coconut oil

25ml/1fl oz almond oil

30ml/2 tbsp ground almonds

10 drops lavender essential oil

heart-shaped moulds, oiled

OLIVE OIL SOAP

Since the 6th century, French craftsmen have been making soaps from olive oil. The centre of soap production in France was Marseilles. Olive oil is rich in vitamin E, which helps the skin to retain moisture and remain bright and supple. Natural, unscented olive oil soaps are a great choice for cleansing sensitive skin.

1 Grate the soap. Place the grated soap in a double boiler and leave it to soften over low heat. Add all the other ingredients.

2 Stir well, until all the ingredients are evenly mixed and begin to hold together.

3 Press the mixture into oiled moulds and leave to set overnight. Unmould the soaps and they are ready to use.

AN ANCIENT REMEDY

The ancient Greek physician Galen is said to have invented cold cream in the 2nd century AD. He did this by melting one part wax and adding three parts olive oil, then blending in as much water as the mixture would hold.

Marigold and sunflower soap

Make this sunny soap from unscented vegetable glycerine soap, adding nut oils, ground sunflower seeds and dried marigold petals.

INGREDIENTS

Makes about 4 soaps

175g/6oz vegetable glycerine soap

25ml/1fl oz coconut oil

25ml/1fl oz almond oil

30ml/2tbsp finely ground sunflower seeds

15ml/1tbsp dried marigold petals

10 drops bergamot essential oil

heart-shaped moulds, oiled

Make these soaps as for the olive oil and lavender soap. Use the same heart-shaped moulds, or try other shapes, if you like.

OILS FOR SOFT SKIN

From head to toe, different vegetable oils and essential oils can be used to make lotions, scrubs and creams to moisturize and condition the skin, leaving it soft and supple.

Soothing body polish

To soothe and polish rough skin, use lavender oil with bergamot, neroli or orange oil for a combination of relaxing and warming aromas.

INGREDIENTS

15ml/1 tbsp almond, avocado or
  macadamia oil
60ml/4 tbsp ground rice
lavender oil
bergamot, neroli or orange essential oil
mild body wash, to use

1 Place a little of the almond, avocado or macadamia oil in a small bowl.

2 Add enough ground rice to make a paste, and a few drops of the essential oils.

3 Before taking a shower, add a small amount of mild body wash and use on any areas of rough skin.

Beauty tip

Scrub from the middle outwards – stomach and down, then up and around the shoulders and arms.

Lavender body lotion

This creamy lotion is perfect for treating dry skin in winter. Borax is the salt of boric acid, known as sodium borate, sodium tetraborate or disodium tetraborate.

INGREDIENTS

1.5ml/^{1}/4 tsp borax
5ml/1 tsp white beeswax
5ml/1 tsp lanolin
30ml/2 tbsp petroleum jelly
25ml/5 tsp plum seed oil
20ml/4 tsp cold pressed sunflower oil
20 drops lavender oil

1 Dissolve the borax in 30ml/2 tbsp boiled water. Melt the beeswax, lanolin and petroleum jelly with the plum seed and sunflower oil in a double boiler.

2 Remove from the heat once the wax has melted and stir well to blend. Add the borax solution while whisking. The lotion will turn white and thicken.

3 Whisk until cool Stir in the lavender oil. Pour into a glass bottle and store in a cool, dark place. Use within 1 year.

Traditional cold cream

This traditional, rose-scented cold cream has a light texture, which is quickly absorbed by the skin, leaving it feeling soft and pampered.

INGREDIENTS

50g/2oz white beeswax
120ml/4fl oz almond oil
50ml/2fl oz rosewater
2.5ml/^{1}/2 tsp borax
120ml/4fl oz bottled spring water, heated

1 Melt the beeswax over a pan of hot water. Off the heat, whisk in the almond oil. Warm the rose water in a pan. Add the borax. Stir until it has dissolved.

2 Add the rosewater mixture to the hot spring water. Whisk into the melted wax and oil. It will start to emulsify, turning white and creamy. Whisk as the mixture cools, to ensure an even texture.

3 Spoon the cream into a jar and seal when it has completely cooled. Use within 6 months.

Coconut and orangeflower body lotion

This creamy preparation is wonderfully nourishing for dry skin. Wheatgerm oil is rich in vitamin E, an antioxidant that protects skin cells against premature ageing.

Olive oil body scrub

INGREDIENTS

120ml/4 fl oz/¹/₂ cup olive oil

30ml/2tbsp sea salt

Mix the ingredients together in a small bowl. Take a shower and, while your skin is still wet, rub the mixture over your body (avoiding any areas of broken skin). Rinse off thoroughly.

INGREDIENTS

50g/2 oz coconut oil

60ml/4 tbsp sunflower oil

10ml/2 tsp wheatgerm oil

10 drops orangeflower essence or 5 drops neroli essential oil

1 Melt the coconut oil in a bowl over gently simmering water. Stir in the sunflower and wheatgerm oils.

2 Leave to cool, then add the fragrance and pour into a jar. The lotion will solidify after several hours. Store in a cool, dark place and use within 6 months.

YOUNGER SKIN

Massage the skin with a mixture of equal parts olive oil and lemon juice to help prevent wrinkles forming.

INEXPENSIVE BODY LOTION

Instead of expensive body lotions, use aqueous cream BP, which is mild and unscented. (It is often recommended by pharmacists and medics for soothing irritated skins and mild rashes.) Transfer a little of the cream (it is usually sold in large pots) to a small sterilized pot and mix in a few drops of essential oil. Prepare a calming mixture for evening and an invigorating one for day. Massage a little macadamia or avocado oil over the skin first, then rub in the cream, which will help the skin to absorb the oil.

OILS FOR HAND TREATMENTS

Oils soften and moisturize the skin and improve the condition of the nails and cuticles. Sweet almond and macadamia are good for their light aromas but olive oil is also very effective.

Winter hand cream

This is a very nourishing cream incorporating patchouli oil, which is a particularly good healer of cracked and chapped skin. Follow the traditional treatment for sore hands by covering them in a generous layer of cream last thing at night and then pulling on a pair of soft cotton gloves. Your hands will have absorbed the cream by morning.

INGREDIENTS

75g/3oz unscented, hard white soap
115g/4oz beeswax
45ml/3 tbsp glycerine
150ml/^{1}/4 pint/^{2}/3 cup almond oil
45ml/3 tbsp rose water
25 drops patchouli oil

1 Grate the unscented soap and place it in a bowl. Pour over 90ml/6 tbsp boiling water and stir until smooth.
2 Combine the beeswax, glycerine,

almond oil and rose water in a double boiler then melt over a gentle heat.

3 Remove from the heat and gradually whisk in the soap mixture. Keep whisking as the mixture cools and thickens.

4 Stir in the patchouli oil and pour into a jar. Store in a cool, dark place. use within 1 year.

Natural nail polish

After trimming and tidying your nails, use oils to condition them.

Soak nails in a little olive oil for 5 minutes, then exfoliate the hands with an oil-based face scrub. Rinse, then polish nails with two or three grades of polishing board and leather. Use a little macadamia or sweet almond oil with the fine grade polisher.

HANDY TIPS

• Massage a little oil and cream into the hands before starting gardening or any other work that is hard on the hands. Tea tree oil is good to massage in with some hand cream before wearing rubber gloves or gardening gloves.

• Rub a little olive oil into your fingernails each night before you go to bed, to stop them chipping and flaking.

Hand cream plus

Using hand cream with oils will promote the absorption of the oil.

Massage a little macadamia nut oil into the hands before using your usual hand cream. Add a drop of essential oil to each hand last thing at night or before relaxing.

OILS FOR FOOTCARE

The moisturizing properties of oil can also be used on the feet, where the skin (especially on the heel area or the side of the big toe) can be dry or rough.

Easy oil pedicure

Regularly using a refreshing peppermint, eucalyptus and lemon scrub on the feet will keep them soft. Using crushed rice or sea salt will remove any rough or hard skin.

FOOT HEALTH

• Anti-fungal and antiseptic oils are excellent for preventing foot odours or worse problems, and also for helping to treat them. Tea tree oil and vinegar are both helpful and they can be combined with olive oil to make a soothing and cleansing rub. Apply generously to the feet and wrap in clear film (plastic wrap), then leave for about 20 minutes to soak in.

• Alternatively, use this combination of ingredients in a foot soak.

• Add tea tree oil to base oil for a massage or to aqueous cream BP with a little peppermint for a lotion. Always thoroughly dry between the toes.

• To avoid picking up infections, wear flip-flops when using communal swimming pools, showers and other wet areas in public facilities.

INGREDIENTS

5ml/1 tsp peppermint oil

5ml/1 tsp eucalytpus oil

5ml/1 tsp lemon oil

15ml/1 tbsp olive oil

60ml/4 tbsp crushed rice or sea salt

100ml/3$\frac{1}{2}$fl oz/scant 1 cup avocado, almond or macadamia oil

1 Start by filing nails to length with an emery board before softening up the skin.

2 Soak the feet in hot soapy water with a few drops of the peppermint oil added.

3 Combine the crushed rice or sea salt with a little olive oil and a drop each of the peppermint, eucalyptus and lemon oils.

4 Use to scrub the feet, concentrating on any areas of rough skin. Wash off the scrub, then rinse the feet in warm water mixed with a drop of peppermint oil.

5 To finish, massage avocado, almond or macadamia oil into the nails and cuticles.

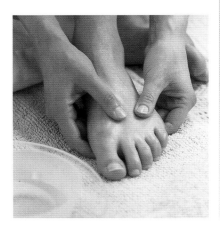

Foot pack

Applying a rejuvenating pack to run-down feet works wonders.

Combine beneficial fruit or vegetable purées with a nourishing oil, such as almond, avocado or olive oil. Wrap the feet in cling film (plastic wrap), enclose in a warm towel and leave to work wonders for 30 minutes.

Foot lotions

Use soothing essential oils on tired feet.

Add a few drops of peppermint or eucalyptus oil to a bottle of body lotion – aqueous cream BP is suitable or use a standard body lotion. For a cooling treat for tired feet, smoothe a little eucalyptus oil into the feet, then massage in witch hazel gel.

OILS FOR HAIRCARE

Essential oils are very good for the hair – rosemary promotes great hair condition, especially for darker hair colours. Camomile is used in treatments for fair hair and to help balance greasy hair.

Essential oil shampoo

Mix up small quantities of shampoo or conditioner using a favourite essential oil, such as rose, lavender, bergamot (good for oily hair) or palmarosa.

INGREDIENTS

shampoo or conditioner

2.5ml/^{1}/$_2$ tsp essential oil

1 Pour shampoo or conditioner into small travel bottles (or rinsed-out hotel bottles) and add a few drops of oil.

2 Mix with a plastic or stainless steel swizzle stick or skewer.

Variation

Make all-in-one shampoo and conditioner by pouring a favourite shampoo into a small bottle. Add a thin layer of conditioner on top of the shampoo and a few drops of rosemary and palmarosa oil (or another combination of favourite oils).

Dry hair conditioning treatment

Egg yolk is naturally rich in vitamins and when mixed with olive oil will work wonders to improve the health and appearance of dry lifeless hair.

INGREDIENTS

1 egg yolk

120ml/4fl oz/^{1}/$_2$ cup olive oil

2 drops lavender essential oil

1 Beat the egg yolk and gradually add the olive oil, then the lavender oil.

2 Massage the egg and oil mixture into your hair and scalp.

3 Soak a towel in hot water, wring it out so that it is damp rather than wet and wrap it around your hair.

4 Leave the treatment in your hair for at least an hour before shampooing, rinsing thoroughly.

Essential oil hair rinse

Mix a favourite essential oil with cider vinegar and rose or orange flower water to remove product build-up from the hair. Try jojoba oil for damaged hair.

INGREDIENTS

30ml/2 tbsp cider vinegar

90ml/6 tbsp rosewater or orange flower water

2 drops essential oil, such as jojoba

1 Mix one quarter cider vinegar to three quarters rosewater or orange flower water in a small container.

2 Add 2 drops of the chosen essential oil. After shampooing and conditioning as usual, rinse your hair.

3 Sprinkle a little of the oil and vinegar hair rinse over the wet hair and massage thoroughly.

4 Gently rinse off the hair.

Rosemary massage treatment

Massage is excellent for promoting circulation and a healthy scalp. Try this super-conditioning three-stage treatment for regenerating tired hair that is dry and lacking in shine. Rosemary is particularly good for dark hair colours.

INGREDIENTS

mild washing up liquid

6 drops rosemary oil

15 ml/1 tbsp jojoba oil

mild shampoo

good-quality unscented or tea
   tree conditioner

1 Begin by brushing the hair well, working very gently so as not to stretch or break the hair at all.

2 Brush the hair in all directions, first brushing it out to ensure it is not tangled, then turning the head upside down. Finally, brush from first one side, then the other.

3 Use a little mild washing up liquid to wash the hair thoroughly, working up a good lather gently without too much excessive rubbing. Take care to avoid tangling or stretching the hair. Rinse the soap off with warm water.

4 Mix 3 drops of rosemary oil into the jojoba oil and massage into the scalp.

5 Begin by distributing through the hair evenly. Then work the fingertips from the sides (above the ears) up and over the top of the head, crossing in the middle. Next work from front to top and back to top in the same way, crossing the fingers on top of the head.

6 Repeat until the oil is thoroughly distributed, lifting the fingers out and away from the head, and through the hair, each time they cross on top.

7 Wash the hair with a mild shampoo to remove the oil.

8 Mix 2–3 drops rosemary oil into a small amount of the unscented or tea tree conditioner, then gently massage this through the hair. Leave for about 30 seconds before rinsing off.

Honey and oil conditioner

Add softness and shine to dry hair with honey and olive oil.

INGREDIENTS

45ml/3 tbsp clear honey

120ml/4fl oz/$\frac{1}{2}$ cup olive oil

2 drops rosemary essential oil

15ml/1 tbsp cider vinegar

1 Mix the honey with the olive oil and add the rosemary essential oil.

2 Massage the mixture into your hair and wrap a hot, damp towel around it.

3 Leave the conditioner on your hair for at least an hour before shampooing and rinsing thoroughly, adding the cider vinegar to the water for the final rinse to make the hair shine.

Intensive hot oil conditioning

This is great for anyone with a dry scalp or dry or colour-damaged hair.

INGREDIENTS

45ml/3 tbsp olive oil

mild shampoo

30ml/2 tbsp cider vinegar

1 Rinse a cup with boiling water to warm it before pouring in 45 ml/3 tbsp olive oil. Stand the cup in a bowl of boiling water to heat the oil.

2 Place a towel to heat on a radiator. Have a roll of cling film (plastic wrap) ready. Wash the hair with a cleansing shampoo or a little detergent, if necessary, to remove any product build-up.

3 Massage the hot oil through the hair, fingering the scalp all over and working the oil out to the ends of the hair.

4 Work the hair up neatly around the head and wrap it all tightly in clear film. Then wrap in a hot towel and relax for 15–30 minutes.

5 Use a mild shampoo to wash out the oil. Massage 30 ml/2 tbsp cider vinegar through the hair and rinse thoroughly.

Camomile and bergamot wash and rinse

This is a very good treatment for fair hair that has a tendency to be greasy.

INGREDIENTS

250ml/8fl oz/1 cup strong camomile tea

3–4 drops bergamot oil

1 Brew a mug of strong camomile tea, preferably with dried camomile flowers and boiling water. Leave to infuse until completely cold, then strain into a jug (pitcher).

2 Add the bergamot oil to the camomile tea. Gently work shampoo into the scalp as usual. Before creating the usual foam, gradually add about a third of the camomile and bergamot water.

3 Continue washing as usual, rinse and condition. Rinse the conditioner off. Finally, rinse the hair with the remaining camomile and bergamot liquid.

HAIRCARE TIP

Decant favourite everyday shampoo and conditioner into small bottles and add a few drops of lavender oil. The oil will impart a pleasant, relaxing aroma.

Overnight lavender treatment

This is an excellent way of benefiting from lavender oil, which will promote a good night's sleep while calming stressed, dry hair. Begin the treatment 2–3 hours before going to bed to avoid having wet hair on the pillow.

INGREDIENTS
3 drops of lavender oil
30 ml/2 tbsp olive oil

1 Wash the hair and towel it dry, then gently comb out any tangles.

2 Mix 3 drops of lavender oil into 30 ml/2 tbsp olive oil and massage this through the hair, working it into the scalp and out through the hair.

3 Gently comb the oil through and leave the hair to dry naturally, rubbing with a towel occasionally.

4 Protect your pillow with a towel to avoid staining or damage. The following morning, wash off with a mild shampoo, then rinse and condition lightly as usual.

Beauty tip
Lavender oil can be used as an effective treatment against head lice.

Hair rescuer

When regularly exposed to the elements, hair can become dry and unmanageable. Central heating, air conditioning and cold weather can all affect the hair's condition. This is a rich nourishing formula, to help improve the condition of dry and damaged hair.

INGREDIENTS
30ml/2tbsp olive oil
30ml/2tbsp light sesame oil
2 eggs
30ml/2tbsp coconut milk
30ml/2tbsp runny honey
5ml/1tsp coconut oil
blender or food processor

1 Place all of the ingredients together in the blender or food processor and process until smooth.

2 Carefully transfer the treatment to a suitable container.

3 After shampooing as usual, comb the mixture through your hair, ensuring even coverage. Leave in the hair for about 5 minutes and then rinse out with warm water.

4 Keep refrigerated and use within three days.

Perk-up conditioner

This makes a zingy conditioner for morning hair washing.

Pour 50ml/2 fl oz/^{1}/4 cup good-quality mild hair conditioner into a small container. Add 1 drop each of rosemary, tea tree and eucalyptus oil to the conditioner and mix together well.

Coconut oil treatment

Using this treatment once a month will work wonders for your hair and scalp.

Mix 90ml/6tbsp coconut oil with 3 drops rosemary oil, 2 drops tea tree oil and 2 drops lavender oil. Use the oil sparingly on dry hair. Coat the hair rather than saturate it, and gently massage it in. Cover with a hot towel for 20 minutes, then shampoo off.

OILS FOR FACIAL SKINCARE

Use milder oils for the delicate skin on the face. Macadamia nut oil is a good choice, but sweet almond oil, wheatgerm oil, avocado oil and olive oil will all benefit dry skin.

5 Gently pat the face dry with a towel and rinse it all over with a mixture of equal quantities witch hazel and rosewater.

Scrub and soothe session

This is a deep cleansing treat, starting with a scrub, followed with a macadamia and palmarosa mask that rounds off to a gentle polish. Salt is thoroughly cleansing but it is harsh – for sensitive skin use the rice flour scrub.

For the scrub

5ml/1 tsp salt

5ml/1 tsp honey

5ml/1 tsp rolled oats or 5ml/1 tsp rice flour

For the freshener

witch hazel

rosewater

For the smoothing mask

5ml/1 tsp macadamia nut oil

10 ml/2 tsp rice flour

2–3 drops palmarosa oil

To finish

2 drops palmarosa oil

5ml/1 tsp macadamia nut oil

1 Mix 5 ml/1 tsp each of salt, honey and rolled oats. (Alternatively, use the honey, oats and 2.5–5ml/^{1}/$_2$–1 tsp rice flour for sensitive skin.)

2 Add a few drops of water to slacken the salt mixture – do not add much or it will become very runny as too much salt dissolves.

3 Gently apply the mixture to the face, avoiding the eye area. Rub lightly only around the nose and chin area and over the forehead. Avoid the lips and eyes, and do not scrub the delicate skin on the cheeks – simply smooth the mixture over the skin.

4 Leave the mixture on the face for 5 minutes. Then wash off, scrubbing gently around the nose and chin.

6 To make the smoothing mask, mix 5 ml/1 tsp macadamia nut oil with about 10 ml/2 tsp rice flour, mixing in just enough to make a smooth, thin paste. Mix in 2-3 drops palmarosa oil.

7 Spread this over the face avoiding the eye area. Leave for 15–30 minutes. Rinse off the mask with lukewarm water, rubbing gently. Pat dry and rinse with witch hazel and rosewater or freshener as before.

8 Mix 2 drops palmarosa oil into 5 ml/1 tsp macadamia nut oil and smooth a little over the face with the tips of the fingers.

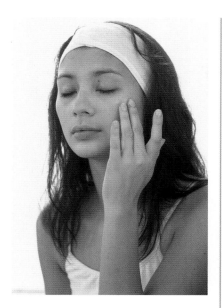

Comfrey and rosewater mask

Herbal face masks tighten the skin, leaving it smooth and fresh. They help to heal blemishes, refine open pores, nourish and soothe. It is best not to use them too often as they can be over-stimulating. This mask is ideal for nourishing dry skin.

INGREDIENTS

6 comfrey leaves
150ml/1/4 pint/2/3 cup boiling water
30ml/2 tbsp fine oatmeal
1 egg yolk
5ml/1 tsp honey
5ml/1 tsp rosewater
5 drops wheatgerm oil
a little milk or yogurt to mix

1 Infuse the comfrey leaves in the boiling water and leave to cool.

2 Strain the infusion into a small bowl.

3 Mix 15ml/1 tbsp of this infusion with the other ingredients to make a smooth paste.

4 Apply evenly to the face, avoiding the eye area. Leave for 10–15 minutes, then rinse off with warm water.

Almond oil cleanser

This almond oil cleanser is a traditional mixture of beeswax, almond oil and rosewater. All creams and lotions are emulsions of oils and water and the addition of a tiny amount of borax means that this mixture emulsifies in a moment, forming a silky-smooth, creamy lotion worthy of the best cosmetic houses. Add essential oils, if you like. Rose oil is suitable for all skin types, and frankincense is particularly good for older skin. To use, smooth it on to the skin using a gentle circular movement and remove with damp cotton wool.

INGREDIENTS

25g/1oz white beeswax
150ml/5fl oz almond oil
1.5ml/1/4 tsp borax (available from chemists)
60ml/4tbsp rosewater
2 drops rose or frankincense essential
  oil (optional)

1 Melt the beeswax in a double boiler and whisk in the almond oil.

2 In a pan, add the borax to the rosewater and warm gently, to dissolve.

3 Slowly add the rosewater mixture to the oils, whisking all the time.

4 Add the essential oil, if using. The mixture will quickly emulsify. Whisk until the mixture has a smooth, creamy texture.

5 Pour into the container, leave to cool and replace the lid securely. Store in a cool, dark place and use within 3 months.

Super-rich avocado face mask

Avocado has been valued as a beauty ingredient for thousands of years. It was used in skincare regimes by the Maya, Aztecs and Incas. The fruit is rich in beneficial oils that will brighten and moisturize tired-looking skin.

Gentle almond face scrub

This is a face scrub worthy of Cleopatra, with its luxurious blend of almonds, oatmeal, milk and rose petals. The rose petals should be bought from an herbalist or, if you want to use petals from your garden, be sure that they have not been sprayed with chemicals. The rose petals can be powdered in a pestle and mortar or in an electric coffee grinder. When mixed with almond oil, the scrub will cleanse the face and leave it silky-soft.

INGREDIENTS

45ml/3 tbsp ground almonds

45ml/3 tbsp medium oatmeal

45ml/3 tbsp powdered milk

30ml/2 tbsp powdered rose petals, or rose oil

30ml/2 tbsp almond oil

1 Mix the ingredients together in a bowl and transfer to a glass jar. Stored in a cool, dark place, the scrub will keep for 1 year.

2 Before using, mix a small amount of the scrub with enough of the almond oil to make a soft paste.

3 Gently rub the scrub into the skin, using a circular motion and being careful to avoid the delicate area around the eyes.

4 Rinse off with warm water and pat your face dry.

INGREDIENTS

150ml/^{1}/4 pint/^{2}/3 cup olive oil

1 large avocado

2 Pour enough of the olive oil over the avocado to make a smooth paste and mix together thoroughly.

1 Peel the avocado and remove the stone (pit). Roughly chop the flesh, place in a small bowl and mash with a fork.

3 Apply to the face and leave on the skin for at least 15–20 minutes before rinsing off with warm water.

Avocado and lemon mask

This rich mask is both moisturizing and refreshing when applied to the face.

INGREDIENTS

1 avocado

5ml/1 tsp lemon juice

5ml/1 tsp avocado oil

cornflour (corn starch), optional

1 Carefully peel the avocado and remove the stone (pit).

2 Purée or sieve (strain) the avocado flesh and stir or whisk in 5ml/1 tsp each of lemon juice and avocado oil.

3 If the mixture seems too runny, sprinkle in a little cornflour to thicken it slightly to a spreading paste.

4 Cleanse the skin thoroughly and then spread the mask over your face, avoiding the delicate eye area.

5 Relax for 15–20 minutes before washing off the mask with warm water.

Beauty tips

Do not use face masks on sore or broken skin. If you have sensitive skin, it is advisable to perform a patch test first on a small area of skin.

Cleansing mask for blackheads

Lemon juice contains natural fruit acids that are beneficial for the skin.

INGREDIENTS

30ml/2 tbsp oatmeal

30ml/2 tbsp yogurt

15ml/1 tbsp lemon juice

15ml/1 tbsp olive oil

1 Place all the ingredients in a small bowl and mix well to make a paste,

2 Smooth the mask over the face.

3 Leave for 10 minutes, then rinse off with cool water.

WHICH OILS TO USE?

• The most useful oils for promoting skin regeneration and general good condition are macadamia nut oil and avocado oil.

• Sweet almond oil is mild, and easily absorbed by the skin and so very good to use in home-made face treatments.

• Geranium, lavender, neroli, orange, palmarosa, rose and rosewood essential oils are all ideal for use in many skin potions, including oil mixtures intended as occasional intensive moisturizers.

Facial mask for dry skin

The vitamins in the egg yolk will add moisture to dry skin.

INGREDIENTS

1 egg yolk

15ml/1 tbsp olive oil

5ml/1 tsp lemon juice

1 Beat the egg yolk in a small bowl and add the olive oil gradually.

2 Add a few drops of lemon juice.

3 Smooth the mixture on to your face and leave until dry, then rinse off with warm water.

Moisturizer aromas

Revitalize a pot of cream that is half used and beginning to lose its fragrance.

Add a few drops of a favourite essential oil to the moisturizer or night cream and mix.

OILS IN THE HOME AND GARDEN

The oils used in cooking may be the most familiar, but there are also non-edible oils that are used around the home. From restoring and protecting furniture to cleaning metals, lubricating rusting hinges and taking care of shoes, it is worth trying some of these traditional materials and methods, instead of relying on commercial products. Using simple techniques and mixtures that have been passed down through generations is satisfying and the results are often a brilliant surprise.

Left: Tung oil and walnut oil are among the oils which may be used around the house for treating wood, lubrication, cleaning or polishing.

USING OILS AROUND THE HOME

The culinary vegetable oils can be put to many good uses outside of the kitchen, but non-edible oils are more suited to tough household jobs.

Oils that are produced for household use do not have to be extracted and processed through a food-safe environment or to the same standards that remove natural toxins. While the following may also be found among culinary oils, the hardware store equivalents are not safe for consumption. The edible kitchen versions of these oils can be put to use around the home, but are usually more expensive.

Useful non-edible oils

Linseed oil This is available from hardware suppliers and is a non-edible product that is reserved for household use, particularly for cleaning and protecting wood. It is essential to distinguish between ordinary or raw linseed oil and boiled linseed oil. Raw linseed oil will not dry to the same extent as boiled linseed oil. It will be absorbed but if an excess is applied it will not dry to a fine glaze but will become very sticky.

Boiled linseed oil was traditionally prepared by heating to 210–260°C/ 410–500°F, sometimes with additional ingredients to promote drying. Blown oil is processed by blowing hot air through it rather than heating in the traditional way. Artists' materials include linseed oils with different drying times, to slow down the drying time of the paint, retard it completely or to speed up the drying time.

Olive oil Non-edible grade olive oil is known as 'lampante' and was traditionally used as a lamp oil. It is common sense that using expensive extra virgin or virgin olive oils for household uses such as polishing tables or to prevent a hinge from squeaking is not sensible but ordinary olive oil can certainly be used for polishing and restoring.

Poppy seed oil This is used in artists' painting products and sold for mixing with paint as a slow-drying medium.

Teak oil This is a blend of oils with excellent drying properties. It is used to treat wood, protecting its surface and enhancing its appearance.

Tung oil This is obtained from the seeds of the tung tree, varieties of which are native to China and Japan.

Above: A varnish will form a protective coating on top of wood, but when wood is treated with boiled linseed oil, the oil soaks into the surface, leaving a shiny finish and emphasizing the natural grain of the wood.

Above: Mix cedar oil with beeswax, turpentine and sandalwood essential oil to make a fragrant furniture polish.

Above: 'Lampante' olive oil is useful for cleaning metal objects.

Above: Oils can be used as a treatment for wood both in the home and the garden.

Above: Teak has a rich golden colour. Over time, items of teak furniture will gradually fade. To prolong the natural colour, the wood should be sanded and treated with teak oil.

It is also referred to as wood oil or Chinese or Japanese wood oil. The oil has excellent drying properties and it is used in artists' materials as well as for treating wood. Being non-toxic, pure tung oil is sold for treating kitchen surfaces, including chopping boards. (Always double check the manufacturer's information to ensure that the particular oil is produced to non-toxic standards before applying it to cooking surfaces.)

Walnut oil This is a good drying oil and one that was traditionally used by artists as a paint thinner and brush cleaner. Walnut oil was preferred to linseed oil as it was less likely to crack or craze when drying.

Cedar oil This oil is made from the cedar tree and has a long history of being used around the home. It formed the base for paints used by the ancient Sumerians. Today, cedar oil is often used in aromatherapy. It can also be used as a floor polish and an insect repellent. A little cedar oil applied to natural cedar furniture will help to renew the lovely woody aroma.

Some uses for oils

There are many ways that these oils can be used around the home and in the garden. Mixtures based on linseed oil can be used to polish and treat wood. Oils are useful in floor polishes and for cleaning and feeding leather, including shoes.

Applying a small amount of oil to an object made from metals such as copper, bronze or brass will provide a barrier between the object and the air and will prevent tarnishing. Using oil in the garden to treat wooden furniture will help it to withstand the elements. A few drops of the right oil will ease stiff hinges, locks or even tools.

A little boiled linseed oil can even be used in the restoration of oil paintings, although great care should be taken to avoid causing damage.

OILS FOR POLISHING AND CLEANING

Before spray polish took over, many people mixed their own potions using oil as a base.
Professional restorers still have their favourite mixtures for restoring antique furniture.

Furniture reviver

Wooden surfaces can become grimy from a combination of dirt and a build-up of spray polish. Use this traditional country recipe to loosen the grime and feed the wood at the same time.

INGREDIENTS

250ml/8fl oz/1 cup malt vinegar
250ml/8fl oz/1 cup pure turpentine
250ml/8fl oz/1 cup raw linseed oil
15ml/1 tbsp sugar

1 Measure all the ingredients into a bottle with a cork or screw top, seal and shake well to mix.

2 Label the bottle clearly.

3 Used over a few weeks, this will gradually remove the layers of polish. Apply with a soft cloth, leave for a few minutes, then wipe off with a second cloth.

Furniture oil

This furniture oil traditionally included benzoin, a resin obtained from an East Indian tree of the same name. It would have been sold in hardware stores a century ago but experimenting today means improvising. The common tincture of benzoin was better known as Friar's Balsam, a product that is still available today from pharmacies or chemist shops. It gives the polish a powerful and refreshing aroma.

INGREDIENTS

200ml/7fl oz/scant 1 cup vinegar
200ml/7fl oz/scant 1 cup boiled linseed oil
25g/1oz benzoin or Friar's Balsam

1 Pour the vinegar, boiled linseed oil and benzoin into an airtight jar.

2 Screw the lid on tight and shake until the mixture is thoroughly combined. Apply to furniture with a soft cloth.

Linseed and shellac polish

Shellac is sold as dry flakes or dissolved in alcohol as liquid shellac varnish. It is a traditional wood finish.

Warm 150ml/^{1}/4 pint/^{2}/3 cup boiled linseed oil and 7g/^{1}/3oz shellac varnish in a small bowl placed over hot water. Stir well to combine. Remove from the heat and allow to cool before using to polish wooden furniture.

TEST PATCH

Furniture polishes should always be tested on a small area first. Absorbent surfaces will change colour when treated with oil-based mixtures. Leather and wood will darken and marks that were insignificant on a dried-out surface may become more prominent. These types of mixtures are ideal for restoring old pieces of furniture, especially less-expensive items that look rather sad and neglected. They are useful for raw wood but not for use on varnished or lacquered modern furniture.

Beeswax and turpentine polish

This is a very simple polish to make and the addition of wood oils will give it an attractive resinous fragrance. Lemon or lavender essential oil may also be used in this polish. Apply to your furniture using a soft cloth, leave a few minutes to dry, then buff vigorously with a soft duster to achieve a deep, lustrous shine.

INGREDIENTS

75g/3oz natural beeswax

200ml/7fl oz/³/4 cup pure turpentine

20 drops cedar oil

10 drops sandalwood oil

1 Grate the beeswax coarsely and place in a screw-top jar.

2 Pour on the turpentine, seal, and leave for a week, stirring occasionally until the mixture becomes a smooth cream.

3 Add the essential oils and mix them in well. The polish is then ready to use.

RENOVATING TIPS

Oil mixtures are ideal for reviving dirty items of furniture, removing years of grime and restoring a shine to old polished surfaces. Use a soft, clean, lint-free cloth to work in the chosen mixture. Work with the grain of the wood, rubbing in the oil or polish along it, not across it, then allow it to soak in or dry before polishing off later. The surface may be rubbed with fine sandpaper and a little boiled linseed oil after the first polish, then left to dry and wiped with a cloth lightly moistened with vinegar before applying more of the oil-based polish.

On untreated wood, some of the best satin finishes that bring out the beauty of light woods are built up over time by cleaning occasionally with a simple mixture that soaks in. Unlike spray polishes that leave a film on the surface, oil-based cleaners feed the wood.

When working on old or damaged furniture, first inspect for woodworm attack and treat the wood accordingly if needed. Then repair any structural damage or wear. Remove surface damage by cleaning and sanding, taking care not to inflict further damage. Old polish or lacquer should be removed by scraping, rubbing down with emery paper and cleaning with white vinegar to provide a smooth, clean surface.

When you are preparing any potentially toxic (and flammable) mixtures, use containers that will not be used for food in future. Heat mixtures carefully, using low heat and taking particular care to avoid direct contact with a naked flame — a heatproof container over a pan of hot water is usually the best way. Always work in a well-ventilated place when handling mixtures that produce strong odours.

Removing marks from furniture

This is a traditional country remedy for cleaning and protecting furniture. The oily mixture will form a light barrier on the surface of the wood, helping to keep it relatively unmarked by fingers. Remember to test the polish on a small area that is out of sight before using on precious items of furniture.

INGREDIENTS

2.4 litres/4 pints water
40g/1¹/₂oz soap flakes
15ml/1 tbsp olive oil

1 Place the water in a large bowl and add the soap flakes.

2 Mix thoroughly with a wooden spoon until the soap flakes have completely dissolved.

3 Add the olive oil and stir until thoroughly combined.

4 Dip a clean cloth into the mixture, wring out any excess liquid, and carefully wipe the piece of furniture all over with the mixture.

5 Dry the furniture thoroughly with a clean, dry cloth.

Non-slip polish for floors

This traditional mixture was intended to be non-slip but it is worth remembering that any polished floors are slippery unless they are made from contemporary products and finishes. There are many variations on the mixture of the popular polishing materials of oil, vinegar and turpentine.

INGREDIENTS

475ml/16fl oz/2 cups white vinegar
475ml/16fl oz/2 cups turpentine
475ml/16fl oz/2 cups boiled linseed oil

1 Place the vinegar, turpentine and boiled linseed oil in an airtight jar.

2 Shake well until the mixture is thoroughly combined.

3 Rub the mixture over the floor with a cloth and leave to dry. Do not buff the floor – this will make it slippery

CLEANING STAINLESS STEEL

Use a small amount of olive oil on a cloth to polish stainless steel surfaces – this will leave them shiny and streak-free. This also works on stainless steel sinks in the kitchen.

Polish for fine scratches

Use this method for minor scratches, otherwise professional polishing is best.

Rub a little walnut oil over a fine scratch, preferably as soon as it is noticed. This will help to restore the colour to the wood and make scratches less noticeable.

Olive oil polish

This is a very simple polish. Use it to clean up pieces of wooden furniture.

Pour equal parts of turpentine and olive oil into a jar with an airtight lid and shake until thoroughly combined. Use to clean and polish furniture, rubbing it over evenly following the grain of the wood. Then polish off with a clean cloth.

Cleaning linoleum

Linoleum (or 'lino') flooring was traditionally cleaned with linseed oil, which restored the colour, protected the surface and kept the flooring in good condition.

Use boiled linseed oil and rub it in well, then polish it off. The result will be a gleaming floor; however, it will also be extremely slippery! Alternatively, mix equal parts of turpentine and olive oil in a small jar, then add a little warm milk and shake well. Use this to clean linoleum or oilcloth.

WEIGHING AND MEASURING

When using mixtures that are toxic, it is important to make sure they do not contaminate equipment that will be used for cooking. Keep a separate mixing spoon, bowl and jug. Pack them away in a separate place (not in the kitchen) after use.

Airtight jars are ideal for the majority of mixing, and precision is not usually vital. Some of the mixtures suggest quantities but it is often a good idea to mix a little in a small jar, judging the proportions by eye, to experiment. If necessary, use electronic kitchen scales on which the bowl or jar can stand.

Cleaning and feeding leather

Olive or other vegetable oil can be used to ease surface dirt off leather. This works well on old briefcases or handbags, suitcases and similar items.

INGREDIENTS

120ml/4 fl oz/$\frac{1}{2}$ cup olive oil

120ml/4 fl oz/$\frac{1}{2}$ cup white vinegar

boiled linseed oil

1 Use a fairly generous amount of oil on a clean, lint-free cloth to gently rub off dirt and feed the leather.

2 Then rub the leather with a little white vinegar on another clean cloth. Leave to dry before finally treating with a mixture of equal parts olive oil and white vinegar.

3 Polish off with a clean cloth.

4 Alternatively, leather upholstery can be cleaned and its texture preserved by rubbing occasionally with a mixture of one part vinegar to two parts boiled linseed oil. The mixture should be used very sparingly.

OILS FOR CLEANING METALS

A little oil can be used on polishing cloths for metals such as copper, bronze or brass, to form a light barrier between the metal and air, and help to minimize tarnishing.

Copper polish

Soft metals such as copper need very careful cleaning. Applying a thin coating of olive oil after cleaning will protect the metal and prevent tarnishing.

Bronze cleaner

Use sweet almond oil plus linseed oil to clean bronze items.

Rub the metal thoroughly with sweet almond oil. Polish off the oil with a clean cloth. Use a small soft brush to clean intricate areas. For really hard-to-clean metal, linseed oil can be used and cleaned off first. Then polish the bronze with sweet almond oil and rub with a chamois leather.

INGREDIENTS

a pinch of salt

1 lemon

15ml/1 tbsp olive oil

1 Sprinkle a pinch of salt on to a saucer. Cut the lemon in half. Take one half and dip the cut surface in the salt. Rub the salty lemon over the copper.

2 Rinse the item in clean hot water and dry immediately with a clean cloth.

3 Place a little olive oil on a clean cloth and rub this over the item before finishing with a clean cloth. To maintain a shine, give the metal an occasional rub with a cloth moistened with olive oil and then polish it off.

Brass cleaner

Use oil on brass to create a barrier between the brass and the air.

After polishing brass items, rub with a cloth dipped in a little olive oil. This will help to prevent tarnishing as well as making it gleam.

OILS IN THE GARDEN

Oil, particularly linseed oil, is well-known for its wood-preserving properties. These can be put to good use on garden furniture that must be treated in order to withstand the elements.

Outdoor wood preservation

Boiled linseed oil is a brilliant wood preservative, especially for sheds or summerhouses that are south-facing and exposed to summer sun, which makes the wood dry out and shrink. The wood must be completely dry and it is best to do this on a warm, but not hot, day and early on in the day so that the oil has time to dry before evening. Applying the oil in hot blazing sun is not ideal as it does not allow for maximum absorption.

Hardwood furniture treatment

Teak oil is best for hardwood furniture, but boiled linseed oil can be used instead.

Rub furniture down with fine sandpaper to loosen the surface and ensure the oil is absorbed. Treat furniture with teak oil regularly and use a cloth dipped in an oil and vinegar mixture (ordinary vegetable oil) to wipe off dust and dirt rather than soapy water.

INGREDIENTS
75ml/5 tbsp boiled linseed oil

1 Thoroughly rub down the wood, first with a brush to remove any loose dirt, then with sandpaper to remove any residue of old surface treatment. Use a wire brush to rub off tough dirt, taking care not to scratch the wood.

2 Rub again with emery paper and brush off with a soft brush.

3 Brush boiled linseed oil generously into the wood, working with the grain and brushing into, under and around joins. Leave to dry for a day or so and then apply a second coat. Treat wood every year and it will last a lifetime!

Greenhouse treatment

This is good for faded and dried out wood as it will absorb a large quantity of oil.

Mix equal parts boiled linseed oil and sunflower oil, and brush on for a first coat. Thoroughly clean, brush down and rub the wood with sandpaper, then brush off all dust and debris. This will bring the wood back to life. Apply a second coat of undiluted boiled linseed oil. This will dry to protect the wood.

OILS FOR LUBRICATION

Every household should have a handy can of light general purpose oil for simple everyday tasks, such as oiling a squeaky hinge or loosening a stiff lock.

Locks

Padlocks or outdoor locks on garden sheds can become stiff or slightly rusty in winter. Use a little vegetable oil to lubricate them.

Scissors, shears and clippers

Gardening equipment stored in a shed tends to stiffen up over the winter months when it may not be in use.

Apply a little vegetable oil and the tools will be easier to use. Apply to the joints with an old paintbrush.

INGREDIENTS

30nl/2 tbsp vegetable oil

1 feather

1 This is an old-fashioned trick for getting oil into a lock. Pour the oil into a saucer.

2 Dip the feather into the oil before inserting it in the lock.

Variations

Alternatively, dip the end of the key into the oil and then insert it into the lock. Drip oil on a stiff or slightly corroded padlock to loosen it. Once it is open, dip the open end in oil and open and shut the lock a few times to lubricate it.

Stacked glasses

If glasses that have been stacked get stuck together, free them with oil.

Pour a small amount of vegetable oil around the rim of the bottom glass. Gently pull the glasses apart – they should separate with ease.

Stiff screws

This is a good solution for old screws that are not badly corroded but just a little stiff.

Place a little oil in a saucer and use a small paintbrush to brush oil on to the screws. A hint of grape seed oil, which is very light, will go a long way but any vegetable oil will help.

Lubricating a sharpening stone

An inexpensive stone is brilliant for keeping kitchen knives sharp. Use vegetable oil to lubricate the stone.

Buy a double-sided sharpening stone, with one side slightly coarser than the other. Drop a little oil on to the fine sharpening side of the stone to finish off sharpening the knives.

Rings on your fingers

When a ring is rather tight and difficult to remove over the knuckle, a little oil may help.

Reduce any swelling as far as possible by holding under cold water and pat it dry. Trickle a little oil on the finger and the ring should slip over the knuckle. Repeat if necessary, taking care not to inflame the finger.

Door hinges

Use oil to loosen a stiff or squeaky hinge.

To avoid oil going everywhere, use a dropper or dip a cotton pad in oil and squeeze it out over the top of the hinge.

Loosening zips

If a zip becomes stuck, loosen it with oil.

Dip a cotton bud in olive oil and dab on to the teeth of the zip, before gently pulling it open.

Hinged furniture

Use oil to stop hinged seats squeaking.

Occasionally brush a little oil on to furniture that has metal mechanisms, for example folding chairs or swinging garden seats.

CLEANING AN OIL PAINTING

This is a good way of gently removing dust and dirt from an inexpensive painting that has been hanging in a room with an open fire.

4 Rub with a sponge moistened with a little tepid water before drying. Polish with a piece of silk cloth.

5 Finish by polishing the painting with a piece of flannel dampened with the boiled linseed oil.

INGREDIENTS
30ml/2 tbsp boiled linseed oil
1 potato

1 To freshen up an old oil painting, moisten a piece of cotton wool with a little boiled linseed oil and gently rub it over the surface of the painting.

WARNING

These methods should only be used on inexpensive paintings. Do not attempt to clean valuable paintings yourself – these should always be taken to professional art restorers.

2 If it still requires a more thorough clean, first cut the potato in half and dip the cut end into cold water.

3 Shake off excess water and rub lightly over the surface of the painting, then wipe with a damp sponge.

OIL AND PAINT

Oil is the medium used for mixing pigments in manufactured oil paints. Oils are also added by artists to dilute the concentrated paint from the tube. Depending on the type of oil added, the paint sets quickly or slowly. Oils commonly used include linseed, poppyseed, walnut and safflower.

OTHER USES FOR OILS IN THE HOME

Beyond the more obvious wood treatments and lubricating, oil can be put to use in other, more suprising ways, from keeping the rust off screws to removing paint from the skin.

Storing screws

This is an old-fashioned but practical way of storing screws, especially for items that are usually used outdoors rather than in clean paintwork in the house.

Some screws are galvanized but others tend to rust, particularly if they are stored outside, in a shed. Instead of leaving them in packets, put them in plastic containers and add a few drops of oil to each. Shake and cover. This will prevent them from rusting.

Shoe care

Traditionally, oils were used to preserve and protect shoe leather. Linseed oil was used for black shoes, castor oil for tan-coloured shoes and olive oil with a little jet black ink added for black patent shoes.

Rub new black shoes with a little olive oil on a cloth to preserve the finish and make cleaning easier. For a quick shoe polish, mix 2 parts olive oil to one part lemon juice and rub into leather shoes with a clean cloth.

Removing labels

Plastic and glass jars are handy to have around the house. Removing labels can be tricky, but soaking in oil will remove any residue.

Fill a small bowl with vegetable oil. Soak the jar(s) in the oil for a few hours, or overnight. The next day, the label will slide right off, taking any sticky glue residue with it. This method also works well for removing sticky price tags.

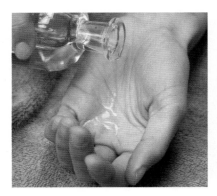

Removing paint

Use oil to clean the skin after painting.

Rub olive oil on to the skin and allow to soak into the skin for about 5 minutes, then rinse off with warm soapy water.

Treating cricket bats

Linseed oil is the traditional coating for the raw willow wood of cricket bats.

Lightly sand the surface of the bat then rub linseed oil into the wood with a clean cloth.

OILCLOTHS AND OILSKINS

Old-fashioned oilcloth was made by treating fabric with boiled linseed oil and allowing the surface to dry.

Oilcloths were the first wipe-clean protective surface covers used on kitchen tables, since they were waterproof and very resilient to damage. Old and tired oilcloth tablecloths can be revived by cleaning with a little boiled linseed oil.

Oilskins, which were traditionally worn by fishermen and sailors, were made in much the same way to produce heavy waterproof garments that were inflexible and rather brittle.

OILS IN THE KITCHEN

Vegetable oils have many roles to play around the house. They can be found in the bathroom cabinet, on the dressing table and in the garden shed, but they are most thoroughly at home in the kitchen. Oils have been used by cooks for many generations in different dishes around the world. They are an important part of a balanced diet and should be included in any store cupboard as a basic ingredient. Frying, marinating, basting, flavouring or simply dressing – this section will help you to explore the many culinary uses of oils.

Left: Vegetable oils can be used in food preparation just as they are, or can be easily flavoured with herbs and spices.

OILS AND DIET

Vegetable oils have many health benefits. They are high in vitamin E and a good source of the fats our bodies need. A variety of oils should be included as part of a balanced diet.

Although low-fat and even fat-free products are popular in today's health-conscious society, we all need to consume a small amount of fat to maintain a healthy and balanced diet. The fat in vegetable oils is a valuable source of energy, and also helps make food more palatable to eat.

Small quantities of polyunsaturated fats are essential for good health and are thought to help reduce the level of cholesterol in the blood. Oils that are a good source of polyunsaturated fats include safflower, corn, sunflower, soya, grape seed and cottonseed oils.

Monounsaturated fats are also though to have the beneficial effect of reducing the blood cholesterol level, and this could explain why in some Mediterranean countries there is such a low incidence of heart disease. Monounsaturated fats are

Above: Olive oil is an essential part of Mediterranean cuisine and a good source of monounsaturated fat, which is thought to play a role in reducing cholesterol levels.

found in nuts such as almonds and hazelnuts, oily fish and avocados. Oils which are a good source of monounsaturated fats include olive, rapeseed (canola), groundnut (peanut) and rice bran oil.

The Mediterranean diet

Olive oil has been used in cooking in Mediterranean countries for centuries. Crete has the highest consumption of olive oil per head in the world: it also has the lowest death rate from cardiovascular disease and the highest proportion of centenarians – a persuasive argument for the merits of the style of eating that is now commonly described as the 'Mediterranean diet'.

Because the Mediterranean region offers little pasture land for raising animals, but does have a lot of sun and sea, its people have traditionally lived on large quantities of ripe fruit and vegetables, herbs, seafood and bread, with relatively little meat and dairy produce.

The Nordic diet

Eating a diet based around the foods local to the Mediterranean region, including olive oil, has long been recommended as a healthy option. However, new research has now shown that the foods traditionally eaten in Scandinavian countries, Norway in particular, may also be among the healthiest.

NUT OILS AND ALLERGIES

Peanut allergies are among the most common food allergies, and the reactions can be severe.

People who suffer from peanut or other nut allergies should avoid using groundnut (peanut) and other nut oils.

The 'Nordic diet' is based around foods which are typically consumed in northern Europe, such as fresh fish and shellfish, leafy green vegetables (such as cabbage and kale) and native berries (such as blueberries and cloudberries). Unlike the Mediterranean countries that use large quantities of olive oil, the most commonly used cooking oil in Scandinavia is rapeseed (canola) oil. Rapeseed oil contains more omega-3 fatty acids than olive oil and is a very good source of vitamin E.

Obesity rates are among the lowest in Scandinavian countries when compared with the rest of Europe, a fact which can only add to the growing popularity of the Nordic diet among the health conscious.

Other health benefits of oil

Cold pressed vegetable oils, (as well as nuts and seeds) are important sources of vitamin E. Known as an important antioxidant, vitamin E plays many roles in ensuring heart and nervous system health, promoting healing and helping to prevent the skin from scarring. Sesame oil is rich in vitamin E and also contains magnesium, copper, calcium, iron and vitamin B6. Linseed oil is said to be an anti-inflammatory, and research is being carried out into the effects of linseed oil as a treatment for sufferers of rheumatoid arthritis. Grape seed oil is believed to lower cholesterol.

Taking small doses of olive oil has been recommended for centuries to aid digestion and reduce the effects of alcohol. Modern studies are still adding to an impressive list of the ways in which eating it regularly can improve your health.

Include olive oil in your diet and it may reduce gastric acidity, and help to protect against ulcers. It helps to prevent constipation, and may even reduce the risk of colon cancer. It stimulates bile secretion and reduces the risk of gallstones.

The antioxidants in extra virgin oil may help to reduce blood pressure. During pregnancy, oleic acid and vitamin E aid the development of the baby's bones, brain and nervous system. It relieves wear and tear on the brain and other organs, reducing the effects of ageing. It also speeds the healing of wounds.

Above: A Scandinavian favourite — marinated fish served with a mustard and dill sauce made with rapeseed oil. The Nordic diet is high in omega-3 fatty acids.

FISH OILS

Fish oils are promoted for their vitamin A and D, and omega-3 fatty acid contents. The fish do not make their own oils but they acquire them through their diet of marine life, especially algae. Currently fish oils are extracted from the waste products from fish and the remaining material is used for fishmeal.

There is much on-going research into ways of introducing omega-3 fatty acids into the human food chain without relying on fish. The natural diet of fish is obviously a source of key interest. Foods are already being supplemented with good fats but the aim for the future is to genetically modify oils as feed for animals, so that they yield meat or eggs that are a source of good fats. Fish oil capsules are often sold in health stores as a dietary supplement.

CULINARY USES OF OILS

Vegetable oils are a staple ingredient in any store cupboard. Oils have been used in the kitchen since ancient times in many different cuisines.

There are three main functions of oil in food preparation: as a cooking medium, to moisten dry foods and make them palatable, and to impart flavour.

Cooking in or with oils

Frying, grilling (broiling), roasting and baking are all methods that use cooking oil (or other fat). Oils used in cooking have to withstand high temperatures without breaking down, smoking and developing unpleasant flavours and unwanted trans fats.

Frying Cooking in or with oil on top of the hob is called frying. Deep-frying immerses the food in oil while shallow-frying cooks the food in a shallow layer of oil. Pan frying is a contemporary term for shallow-frying in minimal oil to prevent the food from sticking;

Above: Oil is drizzled over fish before it is covered and baked in the oven. The oil will impart moisture and flavour to the dish as it cooks.

Above: Stir-frying food in oil is a traditional Eastern cooking technique.

it was adopted instead of 'frying' when the latter became suggestive of unhealthy cooking. Stir-frying is cooking in oil while stirring and tossing the food constantly. Associated with Chinese, Japanese and South-east Asian cooking, stir-frying has acquired a reputation for low-fat cooking. In fact, stir-fried food is not necessarily low in fat but it depends on whether the classic techniques of cooking in stages and adding oil as needed are used, or whether food is simply cooked in one or two stages in the minimum of oil (the modern, Western style of stir-frying). Sautéing, the classic French equivalent of stir-frying, is cooking in oil over high heat while

turning, stirring and tossing the food constantly. For deep-frying, sautéing and stir-frying, the pan and oil have to reach a minimum temperature. Immersing food in oil that is not sufficiently hot makes it greasy.

Grilling (broiling) This form of cooking involves direct heat, either from above, or below (as with a barbecue). Oil is brushed over food before and during grilling to keep it moist and make it crisp.

Roasting This is a method of cooking in the oven, uncovered, without liquid but with fat, such as oil, to baste the food and keep it moist. Traditionally, roasting was done on a rack or a rotating spit, allowing

Above: Richly flavoured olive oil is often served as a dip for fresh bread.

the fat to drip off. Now the food is usually placed in a pan with the fat in the bottom.

Baking This method of oven cooking includes several variations. For example, when breads, cakes, pastries, soufflés and gratins or savoury dishes are baked, the mixture or ingredients are assembled in a tin (pan) or dish. Oil may be included as part of the basic mixture. Some foods are moistened with a little fat or oil before baking, for example fish or vegetables

Above: Add the finishing touch to simple salads with an oil dressing.

may be moistened with butter or oil and cooked in a covered dish, or wrapped in foil, in the oven. Oil is also used to grease dishes or tins to prevent mixtures from sticking during cooking.

Moistening with oil

In addition to acting as a cooking medium, oil is used in several other ways to moisten food. The most basic example is serving oil as a dip for bread – of course, the purpose is to savour full-flavoured oil and good bread but moistening the bread also makes it more palatable.

In the broader sense, oil moistens food when used in cooking and when it is drizzled over as a dressing or used in sauces. Marinating is also used for moistening, tenderizing and flavouring food. Vinegar, wine or fruit juice may be used with oil for marinating, and herbs, spices or other flavouring ingredients are usually added. Marinating can take place over several hours or days in the refrigerator or for a short period at room temperature depending on the main ingredients and the results required.

Above: Oils are used as a flavouring ingredient in many sauces and dips.

Oil as a flavouring ingredient

Many oils have a distinctive flavour. There are many types of olive oil and some are used for their taste, particularly as a dressing for food, to finish a soup or as a dip for bread. Nut oils vary in strength but even the mild macadamia nut oil is quite distinctive. They are particularly useful in dressings and sauces, and they may also be used in sweet dishes, such as chilled desserts.

OIL AS A PRESERVATIVE

Immersing ingredients in oil can be a way of preserving them by excluding air.

However, simply covering foods with oil does not necessarily mean they will keep safely. Commercial processing involves sterilizing the contents to ensure that they are safe. The particular danger is from anaerobic bacteria that thrive in conditions that exclude air.

Botulism is a severe type of food poisoning resulting from eating food contaminated with the *Clostridium botulinum* bacteria; it can be fatal. The contaminated food may look edible.

Bottled foods that have not been properly sterilized and uncooked foods that are preserved in mixtures with low-acidity levels are vulnerable. The high levels of vinegar and sugar in chutneys and pickles together with the cooking process prevents spoilage.

FRYING WITH OILS

From light and crispy deep-fried tempura to speedy stir-fries and sautéed vegetables, the primary way in which oil is used in the kitchen is for frying.

The main point to remember when using oil in cooking is that some types can withstand very high temperatures while others will burn easily, break down, lose their flavour and develop an unpleasant bitter taste.

Recognizing overheating

When oil heats it runs more easily, becoming less viscous and more like a water-based liquid. Tilt a pan when heating oil and notice that it soon runs more easily than when it was cold. When it gets very hot, the oil begins to shimmer slightly – this is the last stage of heating before the oil becomes too hot to use. The shimmering stage is used for some oils that can withstand high heat and for shallow frying some foods. When the oil forms a haze, that means that it has overheated and spoilt.

Checking oil temperature

There are different ways of checking the temperature of oil depending on the cooking method.

Deep-frying Using a sugar thermometer designed for checking high temperatures is the best way of checking the temperature of oil used for deep-frying. As a general rule, the oil should be heated to 180–190°C/350–375°F. Alternatively, use an electric deep-fat fryer. These have thermostats built in.

Cooking a small cube of day-old bread is another method of checking the temperature. When added to oil for deep-frying, the bread should brown in 30–60 seconds.

Shallow-frying or stir-frying When the amount of oil is small, tilt the pan to check the way it flows. When it begins to flow freely it is hot enough for cooking.

Double check by adding a small piece of food first: the food should sizzle immediately but not so harshly that it will burn. Another old-fashioned but successful method is to add a drip of water – just a drip and no more. The fat should spit and sizzle. Adding any more than a drip can be dangerous as the fat will sizzle and spit severely.

DEEP-FRY ALTERNATIVES

Deep-frying is not a very popular cooking method nowadays, and many kitchens do not have a pan suitable for deep-frying. Although a large saucepan can be used, many cooks compromise by frying in about 2.5cm/1in of oil in a sauté pan or wok. The food may not be entirely submerged and therefore it is turned halfway through cooking, but it is effectively deep-fried. Breadcrumb coatings and some batters are suitable for this method. Very light, soft or runny batters may not work on large pieces of food.

SMOKING POINT

When heating oil for cooking, smoking point is the ultimate overheating stage. When this stage is reached, an acrid smell and fine but dark and distinct smoke will be produced. Allowing the oil to heat to this stage not only makes it completely unusable but it is also dangerous as it will easily ignite. Apart from poor flavour and quality, when fats are altered at molecular level this releases free radicals and produces trans fatty acids which should be avoided as far as possible in the diet.

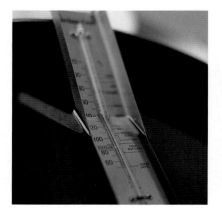

Above: A sugar thermometer can be used to test oil used for deep-frying.

Above: If a cube of bread browns within 60 seconds, the oil is ready to use.

Coating before cooking

The purpose of coating food is to protect the surface of delicate foods from the high temperature of oil to prevent excess drying, disintegration or over cooking. The type of coating depends on the food and frying technique.

Dusting with flour This is useful for shallow frying foods that will not disintegrate in the minimum oil. The oil may form the basis for a sauce and some of the flour from the food may thicken the sauce. Meat, poultry or fish may be floured before frying in a little oil.

Breadcrumbs and batter provide more protection. Egg and breadcrumbs can be used to coat small or large pieces of food, for shallow frying or deep-frying. Batter is not suitable for shallow-frying in the minimum of oil as it will run off but it is used for protecting foods for deep-frying.

Coating with breadcrumbs This is usually referred to as an egg and bread coating. Coating the food with egg enables the breadcrumbs to stick in the first place, then the egg sets quickly and the breadcrumbs form a crisp coating. So that the egg does not slip off, the food is first dusted with flour. Chilling the coated food helps to set the egg and breadcrumb mixture and keep it in place during cooking.

Coating with batter There are several types of batter, some very light and fine, others more robust. The type depends on the food and recipe. The simplest batter is a mixture of self-raising (self-rising) or plain (all-purpose) flour and water, beaten until smooth. The thicker the mixture, the thicker the coating. Thick plain batter is generally unpleasant and a basic batter should be thin. Milk may be used instead of water, as part or all of the amount, although milk tends to make the batter heavy.

Above: Japanese tempura is coated in a very light batter. This lightness is achieved by minimal mixing. The batter protects the food from the very hot oil.

Above: A light coating of flour is often used for food that is shallow fried.

Above: Breadcrumbs will form a crisp coating on food when it is fried.

Eggs may be added, either whole or separated. When separated, the yolks are beaten in first with a little water to make a thick batter, then the whites are whisked until stiff and folded in. This makes a very light batter that rises and becomes thin and crisp during cooking – even if the layer is thicker, because the mixture rises it is far lighter in texture. Fizzy beer is another classic ingredient for making a light batter.

Some batters are prepared by minimum mixing, typically Japanese tempura batter. This very light batter is used to coat pieces of seafood, chopped vegetables and other ingredients, providing a very thin, crisp and 'open' batter. The mixing is so minimal that having tiny pockets of flour is expected – these should not be beaten out as this will give a heavier result.

Perfect deep-frying

Oil used for deep-frying should be able to withstand high temperatures. The best choices are groundnut (peanut), sunflower and blended vegetable oil intended for frying. Olive oil is not used for deep-frying (although there are some exceptions in esoteric recipes).

In this method of frying, the aim is to produce a crisp result that is dry, with the minimum retention of oil at the end of cooking. The oil does not usually flavour the food, but there are exceptions, for example when cooking in corn oil.

A wide variety of foods can be deep-fried as long as it cooks quickly and is tender. The majority of foods need to be protected by an outer coating before deep-frying to prevent the surface from drying out, to retain moisture and to prevent mixtures from disintegrating.

Successful shallow-frying

When using oil for shallow-frying, it is important to differentiate between cooking in oil simply for the 'fried' result and cooking in a particular oil for its flavour.

Shallow-frying in a small amount of olive oil or a blend of sunflower and sesame oil will contribute to the flavour of the dish. Shallow-frying in blended vegetable oil, grapeseed oil or refined groundnut (peanut) oil will produce a 'fried' flavour but without any characteristics of the oil.

Frying for texture When shallow-frying in bland oil, slightly more oil may be used (a thin coating all over the bottom of the pan) and the oil should be well heated before the food is added. The technique is the same as for deep-frying, except that the food has to be turned halfway through cooking. For this type of frying a steady temperature is maintained.

Frying for flavour When the oil is used as the base for a sauce or dressing to serve with the food, the type and amount is important; maintaining a steady cooking temperature may also be vital to cooking other items as well as the main ingredient. For example, fish may be fried in olive oil, then tomatoes, olives and spring onions may be fried in the same oil to be served with the fish.

Above: A small amount of vegetable, groundnut or grapeseed oil is used when shallow-frying food for texture. The oil should just coat the bottom of the pan.

TIPS FOR DEEP-FRYING WITH OIL

A deep pan should be used for frying a significant batch of food or large pieces. Small pieces or a few small items can be cooked in a 'puddle' of oil in the bottom of a wok or small pan but this is not safe for larger quantities. The pan should be one-third to half full of oil before heating and no more. When food is added the level will rise and the hot oil will bubble and froth up. If is boils over the edge of the pan it will easily catch fire.

When the food is added to the oil, the temperature drops. If too large a batch of food is added the oil will not reheat quickly and the food may become soggy and oily. Add small batches for quick, clean cooking.

Turn the food, if necessary, for even cooking. Some foods float and the top may not cook as quickly as the underside.

Always drain food thoroughly. A draining basket in a deep-frying pan allows batches of small items to be lowered into the hot oil and removed all at once. Alternatively, a draining spoon can be used. Let the oil drip off, then place the food on double-thick paper towels to absorb excess oil.

When cooking in several batches, always remove debris from the oil before adding fresh food as the small pieces of food will burn.

Reheat the oil between batches, taking care that it doesn't overheat.

Do not reuse oil more than once or twice – depending on the amount of food cooked, the type, temperature and number of batches. Strongly flavoured foods will taint the oil.

The cooking temperature may be far lower in this type of dish and the overall method more gentle to retain the flavour of the oil that will be scraped out of the pan with the juices and served as an accompaniment.

Slow shallow-frying When cooking thicker food items in shallow oil, it is a good idea to start at a higher temperature and turn the food to brown both sides, then reduce the heat and cook more slowly to allow time for the food to cook through. When using this technique, the flavour of the oil will be imparted to the food and the juices from the food will mix with the oil. The type and amount of oil is important, for example, a thin layer of olive oil may be used or a mixture of oil and butter. The pan juices may be seasoned and sharpened with lemon juice or balsamic (or other) vinegar and served with the food.

Speedy shallow-frying When shallow frying ingredients that cook quickly, preheating the pan brings the oil up to temperature quickly. The disadvantage is that this method is fierce: the oil heats rapidly when added to a very hot pan and

Above: Use a flat-bottomed sauté pan for ingredients that do not take long to cook, such as mushrooms.

it can overheat very quickly. This method is not suitable for non-stick coated pans that should not be heated empty.

Classic sautéing

A sauté pan has a flat bottom and slightly deeper sides than the average frying pan. This allows food to be tossed, shaken and turned. This method

Above: Olive oil is not generally suitable for deep-frying, but is used to impart flavour to shallow fried dishes.

is good for firm but tender ingredients that cook quickly, for example medallions (small, fairly thin round slices) of chicken, pork or lamb; boiled potatoes; or mushrooms.

The ingredients should be cut into even-sized pieces that cook quickly. Refined groundnut (peanut) oil or grape seed oil can be heated to a high temperature and they can be combined with butter. The food should be cooked over high heat and turned and tossed regularly until evenly browned.

Quick and healthy stir-frying

Unlike sautéing, stir-frying usually involves adding small amounts of several different types of foods in stages. Foods that take longest to cook are cooked first, then pushed aside to make room for additional ingredients. All the ingredients have to be cut up evenly into pieces that cook quickly before beginning stir-frying. The pan has to be big enough to hold all the food or the ingredients that are cooked first have to be removed and then replaced at a later stage.

Above: Groundnut oil is popular for stir-frying. Ingredients should all be cut to the same size so that they all cook at the same rate. A little sesame oil can be added for flavour.

CULINARY TRICKS WITH OILS

Most cooks will be used to using oil in their cooking, but it is also helpful for other purposes, from seasoning a wok to greasing tins and as a preserving agent.

Greasing tins (pans)

Use a little mild-flavoured vegetable oil such as sunflower oil to grease cake tins, patty tins (muffin pans) and baking trays.

Use a pastry brush or pour a little of the oil on to a paper towel and wipe around the tin. This will stop cakes or muffins sticking to the tins.

Seasoning and cleaning a wok

Carbon steel woks are sold with a coating to protect them from rust. Seasoning will remove this and make the wok non-stick.

1 Wash the wok thoroughly in warm, soapy water to remove any coating. Rinse well, shake off excess water, then place over a low heat to dry.

2 Add a little oil (not olive oil) and, using kitchen paper, wipe the pan to coat it evenly. Take care not to burn yourself. Heat gently for 10 minutes, then wipe off the oil with clean kitchen paper. Don't be alarmed when it blackens – this is natural.

3 Repeat the process for as long as it takes for the paper to come away clean, by which time the wok itself will have darkened. The more it is used, the better the wok's natural non-stick coating will become and the easier it will be to clean.

4 A properly seasoned wok should never be washed with soap. After use, remove any food that sticks to the surface, then wash in hot water. Dry thoroughly by placing the wok over a low heat for a few minutes. Leave to cool, then rub a little oil into the surface.

Making freezer preserves

These can be kept in the freezer for up to 3 months, ready for adding to a salad, for topping a pizza or enriching a casserole.

Drizzle vegetables such as tomatoes, (bell) peppers, chillies or mushrooms with olive oil, cook slowly until concentrated in flavour. Allow to cool, place in a suitable container then cover with oil. Freeze in small portions.

Oiling a chopping board

Wooden chopping boards, or 'butcher's blocks' need oiling to stop them cracking.

Place 30ml/2 tbsp walnut oil in a small dish and warm in the microwave. Using a clean cloth, liberally apply the oil to the board, rubbing in the direction of the grain. Allow the oil to soak in for 2 minutes, then apply a second coat in the same way. After use, the board should be washed in hot soapy water and dried straight away. If anyone who will use the chopping board has a nut allergy, edible mineral oil (sold as butcher's block oil) should be used instead of walnut oil.

Maintaining a barbecue

Use oil on the barbecue before and after cooking to keep it in top condition.

Before using your barbecue, wipe the grill with a little vegetable oil on a pastry brush. This will prevent food from sticking to the grill. To clean the hot plate of a barbecue, brush with vegetable oil and heat up the plate. The hot oil will make it easier to scrape off any hardened burnt food residue with a spatula. Once it is clean, allow the plate to cool, then brush lightly with oil to prevent rust from forming.

Making super-light cakes

Add a small proportion of sunflower oil, increasing the eggs and flour accordingly.

When making plain creamed cakes, for example, place 115g/4oz butter and 175g/6oz /2/3 cup caster (superfine) sugar in a bowl, then trickle in 45 ml/ 3 tbsp sunflower oil. Cream the mixture as usual, adding 5–10 ml/1–2 tsp vanilla extract, then add 3 eggs with a little flour from 200g/7oz/1¾ cups self-raising (self-rising) flour. Fold in the flour and finish as usual – the result is a delicious light sponge cake.

Making crispy baked potatoes

Brush olive oil or rapeseed (canola) oil on to baking potatoes for extra-crispy skin.

Wash and dry your potatoes thoroughly, then brush with oil and add a generous sprinkling of sea salt. Place on a baking tray and cook in the middle of a hot oven at 220°C/425°F/Gas 7 for 1–1½ hours.

Making Mediterranean mash

Transform your mashed potatoes with a little olive oil and Parmesan cheese.

To make a Mediterranean version of classic mashed potatoes, beat in 15ml/ 1 tbsp olive oil, and season with salt and black pepper. Serve sprinkled with finely grated Parmesan cheese.

Making bread

Making your own bread is very satisfying. Use oiled clear film to help it rise.

Once you have made your dough, always cover it with oiled clear film (plastic wrap) to rise. This will prevent the dough from forming a crust. Coating the clear film with oil will stop the dough from sticking to it.

FLAVOURING OILS

Good quality oil can be flavoured with herbs, spices and aromatics to make rich-tasting oils that are perfect for drizzling, making dressings and cooking.

Cayenne and garlic oil

This is a richly flavoured oil with a kick.

Place 10–15ml/2–3 tsp cayenne pepper and 4 crushed garlic cloves into a clean, dry bottle and pour in 300ml/^{1}/2 pint/1^1/4 cups oil. Leave to infuse (steep) for 2 weeks, then strain and bottle. Keep refrigerated and use within 2 months.

Garlic oil

A first choice herb oil ingredient, garlic is thought to ward off infections. Flavouring oil with garlic is a wonderful way to preserve the tastes of summer, as it can be stored in a cool, dry place for 3–6 months.

INGREDIENTS

5–6 large whole cloves of garlic
600ml/1 pint/2^1/2 cups oil

1 First carefully peel each of the whole garlic cloves.

2 Push the cloves into clean, dry bottles. If the cloves are too large for the neck of the bottle, cut them in half lengthways.

3 Fill the bottles to the top with the oil, and cork. Leave for 2 weeks to infuse (steep). Check the flavour and strain the oil through a muslin- (cheesecloth-) lined sieve (strainer).

4 Pour into clean bottles and keep refrigerated for up to 2 months.

WHICH OILS TO USE?

When flavouring your own oils, use a base oil that is not too strongly flavoured, such as groundnut (peanut), sunflower, safflower or a light olive oil. Avoid using cheap, blended vegetable oils. You should always store home-flavoured oils in the refrigerator.

Chilli and tomato oil

Heating oil with chillies intensifies the rich flavour. This tastes great sprinkled over pasta dishes.

INGREDIENTS

150ml/$^1/_4$ pint/$^2/_3$ cup oil

10ml/2 tsp tomato purée (paste)

15ml/1 tbsp dried red chilli flakes

1 Heat the oil in a large pan. When it is very hot, but not smoking, stir in the tomato purée followed by the chilli flakes.

2 Leave the flavoured oil to cool completely in the pan, then pour it into an airtight jar and store in the refrigerator for up to 2 months.

Dried chilli oil

Leave the chillies in the bottle for a pleasant decorative effect.

Add several dried chillies to a bottle of oil and leave to infuse (steep) for about 2 weeks before using. If the flavour is not sufficiently pronounced, leave for another week. Keep refrigerated for up to 2 months.

Chilli spice oil

Mixed spices enrich this chilli oil.

Add 1 peeled and halved garlic clove, 3 dried red chillies, 5ml/1 tsp coriander seeds, 5ml/ 1 tsp allspice berries, 6 black peppercorns, 4 juniper berries and 2 bay leaves to a bottle of oil. Seal tightly and leave in a cool, dark place for 2 weeks. If the flavour is not sufficiently pronounced, leave the oil to infuse (steep) for another week before straining. Keep refrigerated and use within 2 months.

Basil oil

This has a delicious Mediterranean flavour. The flavour of basil is notoriously difficult to preserve when the herb is dried or frozen, but a well-made basil oil retains all the subtle flavour of the fresh herb. If you wish, use oil that has already been infused with garlic.

INGREDIENTS

about 15g/^{1}/2oz/^{1}/2 cup basil leaves
450ml/^{3}/4 pint/scant 2 cups oil

1 Place the basil leaves in a mortar and pestle and begin to bruise the leaves lightly.

2 Gradually drizzle in about 30ml/ 2 tbsp of the oil. Gently combine the oil with the basil.
3 Pour the oil and basil mixture into a

clean, dry glass bottle and top up with the rest of the oil. Cover and store in a cool place for 2–3 weeks.
4 Pour the infused basil oil through a

muslin- (cheesecloth-) lined sieve (strainer). Leave the oil to drip through the muslin without squeezing the leaves. Pour the strained oil into a clean glass bottle, seal and keep chilled for 2–3 weeks.

Mixed herb oil

Follow this method to flavour oil with different herbs and spices. Parsley, sage and thyme are a good combination, but marjoram, rosemary, bay and sage are also good used together, or singly.

Pour 600ml/1 pint/2^1/2 cups oil into a jar and 125g/4^1/4oz mixed chopped fresh herbs, such as parsley, sage and thyme. Cover and allow to stand at room temperature for about a week, no longer. Shake occasionally during that time. Strain off the oil into a bottle and discard the used herbs. Seal the jar carefully. Store in the refrigerator for up to 6 months.

Lemon oil

Drizzle this subtle fruity oil over pasta dishes. It is good with shellfish.

Finely pare the rind from 1 lemon, place on kitchen paper, and leave to dry for 1 day. Add the dried rind to a bottle of oil and leave to infuse (steep) for up to 3 days. Strain the oil into a clean bottle and discard the rind. Keep refrigerated and use within 2 months.

Spring onion (scallion) oil

The subtle flavour of this spring onion oil make it a great choice for a dressing.

Pour the 250ml/8fl oz/1 cup oil into a pan and heat gently. Stir in 8 finely sliced spring onions (scallions) and remove from the heat. Leave to infuse (steep) until completely cool, then pour the oil into a clean jar or bottle and seal tightly. Store in the refrigerator until ready to use, for up to 2 months.

HERB OILS

• Vegetable oils are widely recognized as being a healthier addition to the diet than butter and other saturated animal fats. Flavoured with herbs, they add a new dimension to cooking.

• As well as tasting delicious, flavoured oils take on the medicinal properties of the herbs used, and so are important constituents of many healing remedies and natural products.

• Oils containing fresh herbs and spices can grow the harmful moulds that can cause botulism. The particular risk is from anaerobic bacteria which thrive when air is excluded. To protect against this, it is recommended that the herbs and spices are removed once their flavour has passed to the oil.

• Adding fresh chillies to a herb oil produces a fiery condiment: try dribbling a tiny amount on to pasta dishes for extra flavour.

Marjoram flower oil

Use any flowers that you have an abundance of to make a fragrant flavoured oil. Thyme, rosemary, lavender, mint and basil are all great alternatives to marjoram flowers. Flower oil makes a perfect summer salad dressing.

INGREDIENTS

30–40 marjoram flower clusters, clean dry and free of insects
450ml/3/4 pint/scant 2 cups oil

1 Fill a large, clean, dry jam jar with the flower clusters (do not worry about removing any small leaves).

2 Pour oil into the jar, covering the flowers, making sure that they are all submerged.

3 Cover with a lid and leave in a warm place for 2 weeks, shaking the jar occasionally.

4 Line a small sieve (strainer) with clean muslin (cheesecloth) or a coffee filter bag and position over a jug (pitcher).

5 Carefully strain the oil into the jug, then pour into a clean attractive bottle. Seal and store in the refrigerator for 3–6 months.

OILS IN MARINADES

These strong-tasting mixes are perfect for adding flavour to meat, poultry, fish and vegetables. Most ingredients should be left to marinate for at least 30 minutes.

Moroccan harissa marinade

Coat fish or shellfish in this fiery marinade to create a dish that will transport you to the North African coastline.

In a small bowl, mix together 30ml/
2 tbsp argan oil, 5ml/1 tsp harissa, 5ml/
1 tsp clear honey and season with a pinch of salt.

Red wine and bay marinade

Marinades containing red wine are particularly good for tenderizing tougher cuts of meat such as stewing steak.

Whisk together 150ml/^{1}/4 pint/^{2}/3 cup red wine, 1 chopped garlic clove, 2 torn fresh bay leaves and 45ml/3 tbsp olive oil. Season with black pepper.

Sweet soy marinade

This can be used to marinade sirloin or fillet steak. It will give it a sweet-sour taste and an oriental feel.

Combine 75ml/5 tbsp dark soy sauce with 30ml/2 tbsp sesame oil with 2 crushed garlic cloves, 15ml/1 tbsp honey and 15ml/1tbsp sesame seeds.

Ginger and soy marinade

This Asian-style marinade is perfect for chicken that is going to be stir-fried.

Peel and grate a 2.5cm/1in piece of fresh root ginger and peel and finely chop a large garlic clove. Whisk together 60ml/4 tbsp olive oil with 75ml/5 tbsp dark soy sauce. Season and stir in the ginger and garlic.

Lemon grass and lime marinade

This light and zesty marinade is great with fish and chicken.

Finely chop 1 lemon grass stalk. In a small bowl, whisk together the grated rind and juice of 1 lime with 75ml/5 tbsp groundnut (peanut) oil, salt and black pepper and the lemon grass.

Rosemary and garlic marinade

This herby marinade is ideal for robust fish, lamb and chicken.

Roughly chop the leaves from 3 fresh rosemary sprigs. Finely chop 2 garlic cloves and whisk together with the rosemary, 75ml/5 tbsp rosemary herb oil and the juice of 1 large lemon.

OILS IN DRESSINGS

Oil is the classic dressing ingredient. These dressings are delicious drizzled over salads but are also tasty served with cooked vegetables and simply cooked fish, meat and poultry.

Walnut dressing

Spoon this dressing over new potatoes and garnish with snipped chives and toasted chopped walnuts.

Combine 30ml/2 tbsp walnut oil with 120ml/4fl oz/^{1}/2 cup low-fat fromage frais or yogurt and add 15ml/1 tbsp chopped fresh flat leaf parsley. Season to taste.

Lemon herb dressing

This quick and easy dressing makes an excellent accompaniment to fish and vegetables.

In a small bowl, whisk together 45ml/ 3 tbsp mixed herb oil with the juice of 1 lemon. Stir in a pinch of dried oregano and season to taste.

Orange and tarragon dressing

Serve this fresh, tangy dressing with salads and grilled (broiled) fish. It will add a light fruity taste.

Whisk together the rind and juice of 1 orange with 45ml/3 tbsp olive oil and 15ml/1 tbsp chopped fresh tarragon. Season with salt and plenty of ground black pepper to taste.

Blue cheese and walnut dressing

Serve this rich dressing with a refreshing sliced pear and watercress salad.

Crumble and then mash 25g/1 oz blue cheese into 30ml/2 tbsp walnut oil. Whisk in 15ml/1 tbsp lemon juice to create a thickish mixture. Season to taste with salt and pepper.

Coconut chilli dressing

This Thai-style dressing is scented with creamy coconut and hot chilli.

Mix 15ml/1 tbsp creamed coconut (coconut cream), 45ml/3 tbsp boiling water, 60ml/4 tbsp groundnut (peanut) oil, rind and juice of 1 lime, 1 chopped red chilli, 5ml/1 tsp sugar and 45ml/3 tbsp chopped fresh coriander (cilantro).

Coriander dressing

This simple dressing made with sesame oil has a pleasing bite.

Mix 120ml/4fl oz/^{1}/2 cup lemon juice, 30ml/2 tbsp wholegrain mustard, 250ml/8fl oz/1 cup olive oil, 75ml/5 tbsp sesame oil and 5ml/1 tsp crushed coriander seeds together in a bowl.

OILS IN SAUCES

In these sauce recipes, oil is infused with other stronger flavourings and ingredients.
Adding nut oils to sauces containing fresh nuts will really bring out their flavour.

Olive oil, tomato and herb sauce

This Mediterranean sauce is enriched
with olive oil. Serve with fresh bread.

INGREDIENTS

Makes approx 450ml/³/4 pint/scant 2 cups

15ml/1 tbsp finely chopped shallot

2 garlic cloves, finely chopped

120ml/4fl oz/¹/2 cup olive oil

about 15ml/1 tbsp lemon juice

225g/8oz tomatoes, peeled, seeded and diced

caster (superfine) sugar

15ml/1 tbsp chopped fresh chervil

15ml/1 tbsp snipped fresh chives

30ml/2 tbsp torn fresh basil leaves

salt and ground black pepper

1 Place the shallot, garlic and oil in a pan
over a low heat and allow to infuse (steep)
for 2 minutes. Whisk in 30ml/2 tbsp cold
water and 10ml/2 tsp lemon juice.

2 Remove from the heat and stir in the
tomatoes. Add a pinch of salt, pepper and
caster sugar, then whisk in the chervil
and chives. Stand for 10 minutes. Reheat
until just warm, stir in the basil, and serve.

Energy 773kcal/3186kJ; **Protein** 2.7g;
Carbohydrate 9g, of which sugars 8.5g; **Fat** 81g,
of which saturates 11.4g; **Cholesterol** 0mg;
Calcium 80mg; **Fibre** 4g; **Sodium** 31mg.

Peanut sauce

The exotic flavour of this peanut sauce
perfectly complements Thai food.

INGREDIENTS

Makes approx 300ml/¹/2 pint/1¹/4 cups

30ml/2 tbsp groundnut (peanut) oil

75g/3oz/³/4 cup unsalted peanuts, blanched

2 shallots, chopped

2 garlic cloves, chopped

15ml/1 tbsp chopped fresh root ginger

1–2 green chillies, seeded and thinly sliced

5ml/1 tsp ground coriander

1 lemon grass stalk, tender base only, chopped

5–10ml/1–2 tsp light muscovado (brown) sugar

15ml/1 tbsp dark soy sauce

105–120ml/3–4 fl oz/scant ¹/2 cup canned
    coconut milk

15–30ml/1–2 tbsp Thai fish sauce (nam pla)

15–30ml/1–2 tbsp tamarind purée (paste)

lime juice

salt and ground black pepper

1 Heat the oil in a small pan and fry the
peanuts, stirring frequently, until lightly
browned. Remove the nuts and drain on
kitchen paper. Set aside to cool.

2 Add the shallots, garlic, ginger, most of
the sliced chillies and the ground
coriander to the pan and cook over a low
heat, stirring, for 5 minutes, until the
shallots are softened but not browned.

3 Transfer the spice mixture to a food
processor or blender and add the
peanuts, lemon grass, 5ml/1 tsp of the
sugar, the soy sauce and 105ml/3fl oz of
coconut milk and the fish sauce. Blend
to a smooth sauce. Add more fish sauce,
tamarind purée, seasoning, lime juice
and/or more sugar to taste. Serve.

Energy 339kcal/1399kJ; **Protein** 4.8g;
Carbohydrate 5.1g, of which sugars 2.8g;
Fat 33.4g, of which saturates 3.5g; **Cholesterol**
0mg; **Calcium** 34mg; **Fibre** 1.2g; **Sodium** 32mg.

Chilli sauce

This hot and spicy sauce is made with groundnut oil – serve with Thai food.

INGREDIENTS

Makes approx 475ml/16fl oz/2 cups

10 fresh red chillies

4 garlic cloves

15ml/1 tbsp Thai fish sauce (nam pla)

15ml/1 tbsp grated palm sugar (jaggery) or soft brown sugar

30ml/2 tbsp lime juice

5ml/1 tsp salt

120ml/4fl oz water

45ml/3 tbsp groundnut (peanut) oil

1 Seed and finely chop the red chillies, and peel and finely chop the garlic cloves, then place in a small pan.

2 Add the rest of the ingredients to the pan and slowly bring to the boil.

3 Reduce the heat and simmer the mixture gently for about 15 minutes.

4 Transfer to a food processor or blender and blend until smooth.

5 Allow to cool, then pour the chilli sauce into a small airtight jar or bottle. The sauce can be stored in the refrigerator for up to 1 week.

Energy 425kcal/1760kJ; **Protein** 8.8g; **Carbohydrate** 21.6g, of which sugars 17.4g; **Fat** 34.3g, of which saturates 4g; **Cholesterol** 0mg; **Calcium** 75mg; **Fibre** 0.8g; **Sodium** 2839mg.

Walnut and garlic sauce

The addition of walnut oil to this sauce gives it a really rich nutty taste.

INGREDIENTS

Makes approx 475ml/16fl oz/2 cups

2 x 1cm/$\frac{1}{2}$in slices good white bread, crusts removed

60ml/4 tbsp milk

150g/5oz/1$\frac{1}{4}$ cups shelled walnuts

4 garlic cloves, peeled and chopped

120ml/4fl oz/$\frac{1}{2}$ cup olive oil

15–30ml/1–2 tbsp walnut oil

juice of 1 lemon

salt and ground black pepper

walnut oil for drizzling

1 Soak the slices of white bread in the milk for about 5 minutes, then process with the walnuts and chopped garlic in a food processor or blender, to make a rough paste.

2 Gradually add the olive oil to the paste with the motor still running, until the mixture forms a smooth thick sauce. Blend in the walnut oil.

3 Scoop the sauce into a bowl and squeeze in lemon juice to taste, season with salt and pepper and beat well. Transfer to a serving bowl, and drizzle over a little more walnut oil.

Energy 1906kcal/7870kJ; **Protein** 28.6g; **Carbohydrate** 34.6g, of which sugars 8.3g; **Fat** 184.7g, of which saturates 20.4g; **Cholesterol** 4mg; **Calcium** 272mg; **Fibre** 6.1g; **Sodium** 324mg.

Perfect mayonnaise

Making great mayonnaise is about slowly incorporating oil with egg yolk.

INGREDIENTS

Makes approx 250ml/8fl oz/1 cups

1 fresh egg yolk

5ml/1tsp French mustard

15ml/1 tbsp lemon juice or white wine vinegar

200ml/7fl oz/scant 1 cup oil

salt and ground black pepper

1 Whisk the egg yolk in a mixing bowl with the seasoning, the French mustard and the lemon juice or vinegar.

2 Pour the oil into a jug (pitcher): olive oil makes mayonnaise with a full flavour; for a lighter result, combine sunflower or grape seed and olive oil.

3 Trickle in a little oil and whisk until combined. Continue whisking while adding oil in a slow trickle, until the mixture is pale and starting to thicken.

4 As the mixture thickens, the oil can be added slightly more quickly – in a steady trickle. Keep whisking until you have a smooth, glossy mayonnaise.

5 The mayonnaise can be stored in the refrigerator in an airtight jar for 1 week.

Energy 958kcal/3940kJ; **Protein** 3.3g; **Carbohydrate** 0,5g, of which sugars 0.4g; **Fat** 104.8g, of which saturates 13.1g; **Cholesterol** 202mg; **Calcium** 27mg; **Fibre** 0g; **Sodium** 157mg.

OILS IN PURÉES AND PASTES

These purées and pastes can be served as an accompaniment to bread, or used as a starting point for other dishes, such as spicy curries or tasty pasta dishes.

Aubergine and yogurt purée

This rich and creamy purée is delicious served with fresh white bread.

INGREDIENTS

Makes approx 500ml/17fl oz/generous 2 cups
2 large aubergines (eggplants)
30ml/2 tbsp olive oil, plus extra for drizzling
juice of 1 lemon
2–3 garlic cloves, crushed
225g/8oz/1 cup natural (plain) yogurt
salt and ground black pepper

1 Grill (broil) the aubergines until the skin is charred and the flesh feels soft. Place in a plastic bag for a few minutes. Hold each aubergine by the stalk under cold running water and peel off the skin until you are left with just the flesh. Squeeze the flesh to get rid of excess water and chop to a pulp, discarding the stalks.

2 Put in a bowl with 30ml/2 tbsp oil, the lemon juice and garlic. Beat well to mix, then beat in the yogurt and season. Transfer to a bowl, drizzle with olive oil and serve immediately.

Energy 103kcal/431kJ; **Protein** 4.4g;
Carbohydrate 7.7g, of which sugars 6.4g;
Fat 6.5g, of which saturates 1.2g; **Cholesterol** 1mg; **Calcium** 118mg; **Fibre** 2.3g; **Sodium** 49mg.

Muhammara

Serve this nutty, spicy Middle Eastern purée with flatbreads.

INGREDIENTS

Makes approx 350ml/12fl oz/1^1/2 cups
175g/6oz/1 cup broken shelled walnuts
5ml/1 tsp cumin seeds, dry-roasted and ground
1–2 fresh red chillies, seeded and finely chopped
1–2 garlic cloves
1 slice of day-old bread, sprinkled with water
  and left for a few minutes, then squeezed dry
15–30ml/1–2 tbsp tomato purée (paste)
5–10ml/1–2 tsp sugar
juice of 1 lemon
120ml/4fl oz/^{1}/2 cup sunflower oil

1 Using a mortar and pestle, pound the walnuts with the cumin seeds, red chilli and garlic.

2 Add the soaked bread and pound to a paste, then beat in the tomato purée, sugar and lemon juice.

3 Drizzle in the oil, beating until the paste is thick. Season, and serve.

Energy 339kcal/1399kJ; **Protein** 4.8g;
Carbohydrate 5.1g, of which sugars 2.8g;
Fat 33.4g, of which saturates 3.5g; **Cholesterol** 0mg; **Calcium** 34mg; **Fibre** 1.2g; **Sodium** 32mg.

Curry paste

Cook spices in oil and store in the refrigerator for convenience.

INGREDIENTS

Makes approx 600ml/1 pint/2^1/2 cups
50g/2oz/^{1}/2 cup coriander seeds
60ml/4 tbsp cumin seeds
30ml/2 tbsp fennel seeds
30ml/2 tbsp fenugreek seeds
4 dried red chillies
5 curry leaves
15ml/1 tbsp chilli powder
15ml/1 tbsp ground turmeric
150ml/^{1}/4 pint/^{2}/3 cup wine vinegar
250ml/8fl oz/1 cup sunflower or corn oil

1 Grind the whole spices to a powder. Transfer to a bowl and mix in the remaining ground spices. Add the vinegar and 75ml/5 tbsp water and stir to a paste.

2 Heat the oil and stir-fry the paste for 10 minutes or until the water is absorbed. When the oil rises to the surface it is cooked. Cool, spoon into airtight jars and store for up to 4 weeks in the refrigerator.

Energy 1949kcal/8040kJ; **Protein** 19.2g;
Carbohydrate 45.4g, of which sugars 0g; **Fat** 191.7g, of which saturates 20.7g; **Cholesterol** 0mg; **Calcium** 234mg; **Fibre** 0g; **Sodium** 44mg.

Thai green curry paste

This fragrant spicy mix makes a delicious Thai curry.

INGREDIENTS

Makes approx 475ml/6fl oz/2 cups

12 fresh green chillies, seeded and chopped

15ml/1 tbsp finely grated garlic

10ml/2 tsp finely grated fresh root ginger

30ml/2 tbsp chopped lemon grass

4 kaffir lime leaves, finely snipped

4 red shallots, chopped

15ml/1 tbsp coriander seeds, roasted

50g/2oz coriander (cilantro) leaves, stalks
   and roots, roughly chopped

5ml/1 tsp ground black pepper

rind of 1 lime, finely grated

90ml/6 tbsp sunflower oil

1 Place all the ingredients in a food processor or blender.

2 Blend to a thick paste. Store in the refrigerator for 2 weeks.

Variation

For red curry paste, blend 8 sliced fresh red chillies, 15ml/1 tbsp roasted coriander seeds, 10ml/2 tsp grated fresh root ginger, 30ml/2 tbsp finely chopped lemon grass, 15ml/1 tbsp grated garlic, 3 finely chopped red shallots, the juice of 1/2 lime and 30ml/2 tbsp sunflower oil.

Energy 793kcal/3273kJ; **Protein** 14.7g;
Carbohydrate 27.3g, of which sugars 14.4g;
Fat 70.5g, of which saturates 8.2g; **Cholesterol**
0mg; **Calcium** 253mg; **Fibre** 6.1g; **Sodium** 45mg.

Traditional pesto

Stir this through pasta for a simple meal, or use as a base for other dishes.

INGREDIENTS

Makes approx 350ml/12fl oz/1 1/2 cups

50g/2oz fresh basil leaves

25g/1oz/1/4 cup toasted pine nuts

2 peeled garlic cloves

20ml/4fl oz/1/2 cup olive oil

25g/1oz/1/3 cup freshly grated
   Parmesan cheese

salt and ground black pepper

1 Put the basil leaves in a food processor or blender and blend to a paste with the toasted pine nuts and peeled garlic cloves. With the motor still running, drizzle in the olive oil through the feeder tube until the mixture forms a paste.

2 Spoon the pesto into a bowl and stir in the freshly grated Parmesan cheese. Season to taste with salt and ground black pepper. To store, drizzle a little oil over the top of the pesto and keep in the refrigerator for up to 1 week.

Variation

To make rocket (arugula) pesto, substitute the fresh basil leaves for 50g/2oz fresh rocket leaves and make in the same way as traditional pesto.

Energy 410kcal/1694kJ; **Protein** 14.9g;
Carbohydrate 2.4g, of which sugars 2.1g;
Fat 38g, of which saturates 8g; **Cholesterol** 25mg;
Calcium 403mg; **Fibre** 3g; **Sodium** 289mg.

Balinese spice paste

This spicy paste forms the basis of many Indonesian meat, poultry and fish dishes.

INGREDIENTS

Makes approx 250ml/8fl oz/1 cup

2 shallots, finely chopped

2 garlic cloves, finely chopped

25g/1oz fresh galangal, finely chopped

25g/1oz fresh turmeric, chopped

4 red chillies, seeded and chopped

1 lemon grass stalk, chopped

5ml/1 tsp ground coriander

2.5ml/1/2 tsp ground black pepper

30–45ml/2–3 tbsp palm oil

10ml/2 tsp shrimp paste

1 Using a mortar and pestle, grind the shallots, garlic, galangal, turmeric, chillies and lemon grass to a coarse paste. Beat in the coriander and black pepper.

2 Heat the oil in a small heavy pan, stir in the paste and fry until fragrant and beginning to colour. Stir in the shrimp paste and sugar and continue to fry for 2–3 minutes, until darker in colour. Remove from the heat and leave to cool.

3 Spoon the spice paste into a jar, cover and store in the refrigerator for up to 1 week.

Energy 158kcal/654kJ; **Protein** 6.6g;
Carbohydrate 8.9g, of which sugars 7.5g;
Fat 13.6g, of which saturates 1.3g; **Cholesterol**
25mg; **Calcium** 25mg; **Fibre** 0.8g; **Sodium** 355mg.

OILS IN DIPS

Oil and vinegar are the classic Mediterranean combination for dipping bread into. If you are feeling more adventurous, try mixing oil with other ingredients in one of these great recipes.

Hummus

Olive oil is an essential ingredient in this traditional Middle Eastern dip.

INGREDIENTS

Makes approx 475ml/16fl oz/2 cups
150g/5oz/³/₄ cup dried chickpeas
juice of 2 lemons
2 garlic cloves, sliced
30ml/2 tbsp olive oil, plus extra for drizzling
pinch of cayenne pepper
150ml/¹/₄ pint/²/₃ cup tahini paste
salt and ground black pepper

1 Soak the chickpeas overnight in cold water. Drain, put in a pan and cover with fresh water. Bring to the boil and cook rapidly for 10 minutes, then simmer gently for about 1 hour, until soft. Drain.

2 Process the chickpeas in a food processor to a smooth purée. Add the lemon juice, garlic, oil, cayenne pepper and tahini and blend until creamy. Season to taste. Transfer to a serving dish and drizzle with more olive oil.

Energy 140kcal/586kJ; **Protein** 6.9g; **Carbohydrate** 11.2g, of which sugars 0.4g; **Fat** 7.8g, of which saturates 1.1g; **Cholesterol** 0mg; **Calcium** 97mg; **Fibre** 3.6g; **Sodium** 149mg.

Taramasalata

This Turkish meze favourite is often served as a dip with pitta bread.

INGREDIENTS

Makes approx 400ml/14fl oz/1²/₃ cups
115g/4oz smoked mullet roe
2 garlic cloves, crushed
30ml/2 tbsp grated onion
60ml/4 tbsp olive oil
4 slices white bread, crusts removed
juice of 2 lemons
30ml/2 tbsp water
ground black pepper
warm pitta bread, to serve

1 Place the smoked fish roe, garlic, grated onion, oil, bread and lemon juice in a blender or food processor and process until smooth.

2 Add the water and process again for a few seconds. Pour into a serving bowl, cover with clear film (plastic wrap) and chill for 2 hours before sprinkling with black pepper and serving.

Energy 807kcal/3374kJ; **Protein** 36.2g; **Carbohydrate** 60.6g, of which sugars 5.4g; **Fat** 48.4g, of which saturates 7.1g; **Cholesterol** 380mg; **Calcium** 147mg; **Fibre** 3g; **Sodium** 700mg.

Tapenade

In France, tapenade is spread on bread and eaten as an hors d'oeuvre.

INGREDIENTS

Makes approx 400ml/14fl oz/1²/₃ cups
225g/8oz/2 cups stoned black olives
2 large garlic cloves, peeled
15ml/1 tbsp salted capers, rinsed
6 canned or bottled anchovy fillets, drained
50g/2oz good-quality canned tuna
5ml/1 tsp chopped fresh thyme
30ml/2 tbsp chopped fresh parsley
30–60ml/2–4 tbsp olive oil
a dash of lemon juice
30ml/2 tbsp crème fraîche or fromage frais

1 Process the olives, garlic, capers, anchovies and tuna in a food processor or blender. Blend in the thyme, parsley and enough olive oil to make a paste.

2 Season to taste with pepper and a dash of lemon juice. Stir in the crème fraîche or fromage frais and transfer to a serving bowl.

Energy 157kcal/666kJ; **Protein** 6.3g; **Carbohydrate** 28.4g, of which sugars 1.8g; **Fat** 2.8g, of which saturates 0.6g; **Cholesterol** 37mg; **Calcium** 76mg; **Fibre** 1.5g; **Sodium** 523mg.

Chilli bean dip

This creamy bean dip is best served warm with triangles of toasted pitta bread.

INGREDIENTS

Serves 4

30ml/2 tbsp corn oil

2 garlic cloves, finely chopped

1 onion, finely chopped

2 green chillies, seeded and finely chopped

5–10ml/1–2 tsp hot chilli powder

400g/14oz can kidney beans

75g/3oz mature (sharp) Cheddar cheese, grated

1 red chilli, seeded and cut into strips

salt and ground black pepper

1 Heat the oil in a large frying pan and add the garlic, onion, green chillies and chilli powder. Cook for 5 minutes until the onions are softened and transparent, but not browned.

2 Drain the kidney beans, reserving the liquor. Blend all but 30ml/2 tbsp of the beans to a purée in a food processor. Add to the pan with 30ml/2 tbsp of the reserved liquor. Heat gently, mixing well.

3 Stir in the whole beans and the cheese. Cook for 3 minutes, stirring until the cheese melts. Add salt and pepper to taste. Transfer to a serving bowl. Scatter the red chilli over the top. Serve warm.

Energy 244kcal/1018kJ; **Protein** 12.7g; **Carbohydrate** 20.2g, of which sugars 4.6g; **Fat** 12.6g, of which saturates 4.8g; **Cholesterol** 18mg; **Calcium** 234mg; **Fibre** 7.1g; **Sodium** 538mg.

Chunky cherry tomato salsa

The chilli oil in this refreshing dip gives it that extra fiery edge.

INGREDIENTS

Serves 4

1 cucumber

5ml/1 tsp sea salt

500g/11/4lb cherry tomatoes, quartered

1 lemon

45ml/3 tbsp chilli oil

2.5ml/1/2 tsp dried chilli flakes

30ml/2 tbsp chopped fresh dill

1 garlic clove, finely chopped

1 Trim the ends off the cucumber and cut it into 2.5cm/1 in lengths, then cut each piece lengthways into thin slices.

2 Arrange the slices in a colander and sprinkle with sea salt. Leave for 5 minutes, then wash under cold water and dry with kitchen paper. Place in a bowl with the tomato pieces.

3 Finely grate the lemon rind and place in a small jug (pitcher) with the juice from the lemon, the chilli oil, chilli flakes, dill and garlic. Whisk with a fork.

4 Pour the chilli oil dressing over the tomato and cucumber and toss. Leave to marinate for 2 hours, then serve.

Energy 162kcal/676kJ; **Protein** 4.2g; **Carbohydrate** 16g, of which sugars 13.7g; **Fat** 9.5g, of which saturates 1.4g; **Cholesterol** 0mg; **Calcium** 83mg; **Fibre** 4.5g; **Sodium** 31mg.

Avocado and red pepper salsa

This simple salsa is a fire-and-ice mixture of hot chilli and cooling avocado.

INGREDIENTS

Serves 4

2 ripe avocados

1 red onion

1 red (bell) pepper

4 green chillies

30ml/2 tbsp chopped fresh coriander (cilantro)

20ml/2 tbsp sunflower oil

juice of 1 lemon

salt and ground black pepper

1 Halve and stone the avocados. Scoop out and finely dice the flesh. Finely chop the red onion.

2 Slice the top off the pepper and pull out the central core. Shake out any remaining seeds. Cut the pepper into thin strips and then into fine dice.

3 Halve the chillies, remove their seeds and finely chop the flesh. Mix the chillies, coriander, oil, lemon and salt and pepper to taste.

4 Place the avocado, red onion and pepper in a bowl. Pour in the chilli and coriander dressing and toss the mixture well. Serve immediately.

Energy 216kcal/890kJ; **Protein** 2.4g; **Carbohydrate** 5.8g, of which sugars 4.2g; **Fat** 20.3g, of which saturates 3.8g; **Cholesterol** 0mg; **Calcium** 41mg; **Fibre** 4.1g; **Sodium** 11mg.

OILS IN SOUPS

Whether as a base for rich vegetable soups or as a finishing touch to add flavour, oils are an essential ingredient in these delicious, warming recipes.

White radish and beef soup

The smoky flavours of beef are perfectly complemented by the tanginess of Chinese white radish in this mild and refreshing soup with a slightly sweet edge.

INGREDIENTS

Serves 4

200g/7oz Chinese white radish, peeled

50g/2oz beef, chopped into bitesize cubes

15ml/1 tbsp sesame oil

1/2 leek, sliced

15ml/1 tbsp soy sauce

salt and ground black pepper

1 Slice the white radish and cut the pieces into 2cm/3/4in squares.

2 Heat the sesame oil in a large pan and stir-fry the beef until golden brown. Add the white radish and briefly stir-fry.

3 Add 750ml/11/4 pints/3 cups water. Boil, then simmer, covered, for 7 minutes. Add the leek, soy sauce and seasoning. Simmer for 2 minutes. Serve.

Energy 60kcal/247kJ; **Protein** 3.7g; **Carbohydrate** 2g, of which sugars 1.8g; **Fat** 4.1, of which saturates 1g; **Cholesterol** 7mg; **Calcium** 17mg; **Fibre** 1g; **Sodium** 281mg.

Tomato, ciabatta and basil oil soup

Basil oil captures the delicate flavour of this classic Italian herb and makes a simple tomato soup something special.

INGREDIENTS

Serves 4

45ml/3 tbsp olive oil

1 red onion, chopped

6 garlic cloves, chopped

300ml/1/2 pint/11/4 cups white wine

150ml/1/4 pint/2/3 cup water

12 plum tomatoes, quartered

2 x 400g/14oz cans plum tomatoes

2.5ml/1/2 tsp sugar

1/2 ciabatta loaf

salt and ground black pepper

basil leaves, to garnish

120ml/4fl oz/1/2 cup basil oil

1 Heat the oil in a large pan and cook the onion and garlic for 4–5 minutes until softened.

2 Add the wine, water, fresh and canned tomatoes. Bring to the boil, reduce the heat and cover the pan, then simmer for 3–4 minutes. Add the sugar and season well with salt and black pepper.

3 Break the bread into bitesize pieces and stir into the soup.

4 Ladle the soup into bowls. Garnish with basil and drizzle the basil oil over each portion.

Energy 332kcal/1396kJ; **Protein** 7.8g; **Carbohydrate** 35.4g, of which sugars 16.3g; **Fat** 13.4g, of which saturates 2g; **Cholesterol** 0mg; **Calcium** 98mg; **Fibre** 5g; **Sodium** 306mg.

Crab, coconut and coriander soup

This South American soup combines the distinctive flavours of creamy coconut, palm oil, fragrant coriander and chilli.

INGREDIENTS

Serves 4

30ml/2 tbsp olive oil

1 onion, finely chopped

1 celery stick, finely chopped

2 garlic cloves, crushed

1 fresh chilli, seeded and chopped

1 large tomato, peeled and chopped

45ml/3 tbsp chopped fresh coriander (cilantro)

1 litre/1³/₄ pints/4 cups fresh fish stock

500g/1¼lb crab meat

250ml/8fl oz/1 cup coconut milk

30ml/2 tbsp palm oil

juice of 1 lime

salt

hot chilli oil and lime wedges to serve

1 Heat the olive oil in a pan over a low heat. Stir in the onion and celery and sauté gently for 5 minutes until softened and translucent. Stir in the garlic and chilli and cook for 2 minutes.

2 Add the tomato and half the coriander and increase the heat. Cook, stirring, for 3 minutes, then add the stock. Bring to the boil, then simmer for 5 minutes.

3 Stir the crab, coconut milk and palm oil into the pan and simmer over a very low heat for a further 5 minutes. The consistency should be thick, but not stew-like, so add some water if needed.

4 Stir in the lime juice and remaining coriander, then season with salt to taste. Serve in heated bowls with the chilli oil and lime wedges to serve.

Energy 228kcal/951kJ; **Protein** 23.6g; **Carbohydrate** 5.4g, of which sugars 5g; **Fat** 12.6g, of which saturates 3.7g; **Cholesterol** 90mg; **Calcium** 199mg; **Fibre** 1.1g; **Sodium** 767mg.

Peanut and potato soup

This soup is wonderfully warming. Serve with peanuts for added crunch.

INGREDIENTS

Serves 6

1 onion, finely chopped

60ml/4 tbsp groundnut (peanut) oil

2 garlic cloves, crushed

1 red (bell) pepper, seeded and chopped

250g/9oz potatoes, peeled and diced

2 fresh red chillies, seeded and chopped

200g/7oz canned chopped tomatoes

150g/5oz/1¼ cups unsalted peanuts

1.5 litres/2½ pints/6¼ cups beef stock

salt and ground black pepper

1 Cook the onion in the oil for 5 minutes, until beginning to soften. Add the garlic, pepper, potatoes, chillies and tomatoes. Stir well to coat the vegetables evenly in the oil, cover and cook for 5 minutes.

2 Toast the peanuts in a dry frying pan. Set 30ml/2 tbsp of the peanuts aside. Transfer the remaining peanuts to a food processor and process until finely ground. Add the vegetables and process again until smooth.

3 Return to the pan and stir in the beef stock. Bring to the boil, lower the heat and simmer for 10 minutes. Serve garnished with the remaining peanuts.

Energy 260kcal/1079kJ; **Protein** 8g; **Carbohydrate** 14.7g, of which sugars 6.2g; **Fat** 19.2g, of which saturates 3.6g; **Cholesterol** 0mg; **Calcium** 30mg; **Fibre** 3g; **Sodium** 20mg.

OILS IN SALADS

Oil is the classic dressing ingredient for salads. These recipes often use olive oil, but you could also try a home-flavoured herb oil or a more distinctive nut oil.

Tomato salad with marinated peppers and oregano

The simple combination of olive oil and white wine vinegar is used to dress this classic summer salad.

INGREDIENTS

Serves 2–4

2 marinated (bell) peppers, drained

6 ripe tomatoes, sliced

15ml/1 tbsp chopped fresh oregano

75ml/5 tbsp olive oil

30ml/2 tbsp white wine vinegar

sea salt

Roasted red peppers with feta, capers and preserved lemons

The Mediterranean flavours of this dish cry out for the distinctive taste of argan oil, but you can use olive oil too.

INGREDIENTS

Serves 4

4 red (bell) peppers

200g /7oz feta cheese, crumbled

30–45ml/2–3 tbsp argan oil or olive oil

30ml/2tbsp capers

peel of 1 preserved lemon, cut into
  small pieces

salt (optional)

1 Preheat the grill (broiler) on the hottest setting. Roast the red peppers under the grill, turning frequently, until they soften and their skins begin to blacken.

2 Place the peppers in a plastic bag, seal and leave them to stand for 15 minutes.

3 Peel the peppers, remove the stalks and seeds, then slice the flesh and arrange on a plate.

4 Add the crumbled feta and pour over the argan oil or olive oil.

5 Scatter the capers and preserved lemon over the top and sprinkle with a little salt, if required (this depends on whether the feta is salty or not). Serve with chunks of fresh bread to mop up the delicious, oil-rich juices.

Energy 233kcal/967kJ; **Protein** 9.8g; **Carbohydrate** 12.2g, of which sugars 11.6g; **Fat** 16.4g, of which saturates 7.8g; **Cholesterol** 35mg; **Calcium** 209mg; **Fibre** 3.2g; **Sodium** 730mg.

1 Cut the marinated peppers into strips. Arrange the tomato slices and pepper strips on a serving dish, sprinkle with the oregano and season to taste with sea salt.

2 Whisk together the olive oil and vinegar in a jug (pitcher) and pour over the salad. Serve immediately or cover and chill in the refrigerator until required.

Energy 119kcal/494kJ; **Protein** 1.4g; **Carbohydrate** 6.9g, of which sugars 6.7g; **Fat** 9.7g, of which saturates 1.5g; **Cholesterol** 0mg; **Calcium** 17mg; **Fibre** 2.1g; **Sodium** 12mg.

Bulgur wheat salad

This traditional meze dish of bulgur and tomato is easy to make and delicious. Olive oil adds to the mix of flavours and stops the dish being too dry.

INGREDIENTS

Serves 4–6

175g/6oz/1 cup bulgur wheat, rinsed
   and drained

45–60ml/3–4 tbsp olive oil

juice of 1–2 lemons

30ml/2 tbsp tomato purée (paste)

10ml/2 tsp sugar

1 large or 2 small red onions, cut in half
   lengthways, in half again crossways, and
   sliced along the grain

1–2 fresh red chillies, seeded and
   finely chopped

1 bunch each of fresh mint and flat leaf
   parsley, finely chopped

salt and ground black pepper

a few fresh mint and parsley leaves,
   to garnish

1 Put the bulgur into a wide bowl, pour over enough boiling water to cover it by about 2.5cm/1in, and give it a quick stir.

2 Cover the bowl with a plate or pan lid and leave the bulgur to steam for about 25 minutes, until it has soaked up the water and doubled in quantity.

3 Pour the oil and lemon juice over the bulgur and toss to mix, then add the tomato purée and toss the mixture again until the bulgur is well coated.

4 Add the sugar, onion, red chillies and the herbs. Season with salt and pepper.

5 Serve at room temperature, garnished with a little mint and parsley.

Energy 149kcal/620kJ; **Protein** 3g;
Carbohydrate 21.6g, of which sugars 5.4g;
Fat 6.1g, of which saturates 0.8g; **Cholesterol**
0mg; **Calcium** 54mg; **Fibre** 1.7g; **Sodium** 19mg.

Stilton and walnut salad

The addition of walnut oil in the dressing gives this simple salad a really rich, nutty flavour.

INGREDIENTS

Serves 4

mixed salad leaves

4 fresh figs

115g/4 oz stilton, cut into small chunks

75g/3oz/³⁄₄ cup walnut halves

For the dressing

45ml/3 tbsp walnut oil

juice of 1 lemon

salt and ground black pepper

1 Mix all the dressing ingredients together in a bowl. Whisk briskly until thick and emulsified.

2 Wash and dry the salad leaves, then tear them gently into bitesize pieces. Place in a mixing bowl and toss with the dressing. Transfer to a large serving dish.

3 Cut the figs into quarters and add to the salad leaves. Sprinkle the cheese over, crumbling it slightly. Then sprinkle over the walnuts, breaking them up roughly with your fingers as you go.

Energy 415kcal/1726kJ; **Protein** 10.6g;
Carbohydrate 26.6g, of which sugars 26.4g; **Fat**
30.3g, of which saturates 7.3g; **Cholesterol** 22mg;
Fibre 4.5g; **Sodium** 383mg.

OILS IN DEEP-FRIED SAVOURY DISHES

Use an oil that can withstand high temperatures for deep-frying, such as sunflower or groundnut (peanut) oil. If food is fried in oil that is not hot enough, it will become soggy.

Japanese tempura

In this classic dish from Japan, ingredients such as prawns and vegetables are coated in a light batter and deep-fried very briefly in sunflower or groundnut oil.

INGREDIENTS

Serves 4

50g/2oz/^{1}/$_2$ cup rice flour

100g/3^3/$_4$oz/scant 1 cup plain (all-purpose) flour

2.5ml/^{1}/$_2$ tsp baking powder

2 large (US extra large) egg whites, beaten

350–400ml/12–14fl oz/1^1/$_2$–1^2/$_3$ cup ice-cold soda water (club soda)

sunflower oil, or groundnut (peanut) oil for deep-frying

8 raw tiger prawns (jumbo shrimp), peeled, with tails intact

300g/11oz assorted vegetables, such as sliced sweet potato, broccoli florets, baby spinach leaves, courgette (zucchini) slices, sliced red and yellow (bell) pepper

1 Preheat the oven to low. Mix the flours, baking powder, beaten egg whites and soda water and stir until just combined. (Do not overmix or the batter will become heavy. It should be lumpy.)

2 Pour the oil into a wok to a depth of 7.5cm/3in and heat to 180°C/350°F. Working in batches, dip the prawns and vegetables in the batter, shaking off the excess batter, and deep-fry for 3 minutes, or until golden.

3 Drain the tempura on a wire rack lined with kitchen paper, then transfer to another wire rack (with no paper) and place in the oven to keep warm. Cook the remaining tempura and serve immediately with dipping sauce.

Energy 669kcal/2,775kJ; **Protein** 48.7g; **Carbohydrate** 5g, of which sugars 3.8g; **Fat** 45.9g, of which saturates 11.1g; **Cholesterol** 250mg; **Calcium** 37mg; **Fibre** 0.9g; **Sodium** 196mg.

Prawn and sesame toasts

These little toasts are quickly fried in oil.

INGREDIENTS

Serves 4

225g/8oz peeled raw prawns (shrimp)

15ml/1 tbsp sherry

15ml/1 tbsp soy sauce

30ml/2 tbsp cornflour (cornstarch)

2 egg whites

4 slices white bread, cut into quarters

115g/4oz/^{1}/$_2$ cup sesame seeds

groundnut (peanut) oil, or sunflower oil for deep-frying

1 Process the prawns, sherry, soy sauce and cornflour in a food processor. Whisk the egg whites until stiff. Fold them into the prawn and cornflour mixture.

2 Put the sesame seeds on to a plate. Spread the prawn paste over one side of each bread triangle, then press into the sesame seeds. Heat the oil in a wok to 190°C/375°F. Add the toasts, prawn side down, and deep-fry for 2 minutes, then turn and fry on the other side until golden. Drain on kitchen paper and serve.

Energy 433kcal/1,806kJ; **Protein** 19.1g; **Carbohydrate** 27.7g, of which sugars 1.2g; **Fat** 27.6g, of which saturates 3.6g; **Cholesterol** 110mg; **Calcium** 271mg; **Fibre** 2.7g; **Sodium** 559mg.

Deep-fried wontons

When frying the wontons, make sure
your oil is not too hot – they will scorch.

INGREDIENTS

Serves 4

300g/11oz/1½ cups minced (ground) pork

15ml/1 tbsp light soy sauce

15ml/1 tbsp sesame oil

2.5ml/½ tsp ground black pepper

15ml/1 tbsp cornflour (cornstarch)

16 wonton wrappers

groundnut (peanut) oil or sunflower oil,
  for deep-frying

chilli dipping sauce, to serve

1 Put the minced pork in a bowl. Add
the light soy sauce, sesame oil, ground
black pepper and cornflour. Mix well.

2 Place about 5ml/1 tsp of the mixture
in the centre of a wonton wrapper, bring
the corners together so that they meet
at the top, and pinch the neck to seal. Fill
the remaining wontons in the same way.

3 Heat the groundnut or sunflower oil in
a wok or deep-fryer. Carefully add the
filled wontons, about four or five at a
time, and deep-fry until golden brown.

4 Carefully lift out the cooked wontons
with a slotted spoon, drain on kitchen
paper and keep hot while frying
successive batches. Serve the wontons
hot with chilli dipping sauce.

Energy 326kcal/1357kJ; **Protein** 16.3g;
Carbohydrate 18.3g, of which sugars 0.6g;
Fat 21.3g, of which saturates 4.4g; **Cholesterol**
50mg; **Calcium** 33mg; **Fibre** 0.6g; **Sodium** 319mg.

Chips (French fries)

These chips are fried twice in oil –
once to cook them and then a second
time to brown them and give them a
crispy coating.

INGREDIENTS

Serves 4

groundnut (peanut) or sunflower oil,
  for frying

675g/1½lb potatoes

salt

1 Heat the oil to 150°C/300°F. Peel the
potatoes and cut them into chips (fries)
about 1cm/½in thick. Rinse and dry.

2 Lower a batch of chips into the hot
oil and cook for about 5 minutes or
until tender but not browned.

3 Lift out the chips with a slotted spoon,
place on kitchen paper and leave to cool.

4 Just before serving, increase the
temperature of the oil to 190°C/375°F.
Add the par-cooked chips, in batches.

5 Cook until crisp, then lift out and
drain on kitchen paper.

6 Sprinkle with salt and serve at once.

Energy 403kcal/1689kJ; **Protein** 5.4g;
Carbohydrate 51.5g, of which sugars 2.9g;
Fat 14.5g, of which saturates 6.1g; **Cholesterol**
0mg; **Calcium** 19mg; **Fibre** 3.7g; **Sodium** 59mg.

OILS IN DEEP-FRIED SWEET DISHES

Deep-frying food gives it a crispy finish while allowing the centre of the food to stay soft – which makes it an ideal cooking technique for these sweet dishes.

Cinnamon and apple fritters

Served piping hot with a dollop of crème fraîche, these slices of crisp apple in a fluffy, cinnamon-scented batter will fill the house with a mouth-watering aroma. Preserve the flavour and lightness of the batter by deep-frying the fritters in very hot vegetable oil.

INGREDIENTS

Serves 4–6

3–4 tart eating apples such as Granny Smith

50ml/2fl oz/$^1/_4$ cup Marsala

3 eggs, beaten

125g/4$^1/_4$oz/generous 1 cup plain
(all-purpose) flour

30g/1$^1/_4$oz/generous 2 tbsp golden caster
(superfine) sugar

5ml/1 tsp ground cinnamon

2.5ml/$^1/_2$ tsp salt

vegetable oil, for deep-frying

berries, to decorate

caster (superfine) sugar, for dusting

crème fraîche, to serve

1 Peel, core and slice the apple into 1cm/$^1/_2$in rings. Place the apple slices in a large, shallow bowl and pour over the Marsala, turning to coat them evenly. Cover and set aside to macerate for about 1 hour.

2 Beat together the eggs, flour, sugar, cinnamon and salt in a large bowl until thick and smooth. Drain the apples and set aside, reserving the Marsala. Add enough of the Marsala to the batter to make a coating consistency. Beat until smooth and free of lumps.

3 Add the apples to the batter and stir gently to coat evenly. Heat the oil in a deep-fat fryer or a large, heavy pan to 180°C/350°F.

4 Working in batches, lower the apple rings into the oil and deep-fry for 3–4 minutes until golden. Remove with a slotted spoon and drain on a wire rack placed over crumpled kitchen paper. Divide the fritters among warmed serving plates and decorate with berries. Dust with caster sugar and serve immediately with crème fraîche.

Energy 286kcal/1195kJ; **Protein** 5.4g;
Carbohydrate 29.9g, of which sugars 14g; **Fat** 16g, of which saturates 2.2g; **Cholesterol** 95mg;
Calcium 50mg; **Fibre** 2g; **Sodium** 38mg.

American-style sugared donuts

These irresistible, all-time favourite treats are cooked without yeast for a shorter preparation time and almost instant results.

INGREDIENTS

Serves 4

60g/2¹/₄oz/generous 4 tbsp unsalted butter

60g/2¹/₄oz/generous ¹/₄ cup caster (superfine) sugar

1 large (US extra large) egg, beaten

90ml/6 tbsp buttermilk

225g/8oz/2 cups plain (all-purpose) flour

5ml/1 tsp baking powder

2.5ml/¹/₂ tsp bicarbonate of soda (baking soda)

pinch of salt

finely grated rind of 1 orange

5ml/1 tsp ground cinnamon

pinch of grated nutmeg

vegetable oil, for deep-frying

icing (confectioners') sugar, for dusting

1 Place the butter, sugar, beaten egg and buttermilk in a large bowl.

2 Sift over the flour, baking powder, bicarbonate of soda and salt.

3 Add the orange rind, cinnamon and nutmeg and mix using a wooden spoon until well-mixed and smooth.

4 Turn the mixture out on to a lightly floured work surface. With floured hands, knead the dough for about 5 minutes until it becomes really smooth and slightly soft and elastic but not sticky.

5 Roll out the dough on a lightly floured surface to 1.5cm/²/₃in thick and stamp out rounds using a 5cm/2in cutter. Press a hole through each one with the handle of a wooden spoon to make a ring shape.

6 Pour the oil into a deep fat fryer, wok or a large, heavy pan and heat to 180°C/350°F.

7 Fry the donuts, in batches, for about 5 minutes, or until browned and cooked through. Lift out with a skimmer and drain on a wire rack lined with kitchen paper. Leave to cool slightly, then dust with sifted icing sugar and serve warm.

Energy 192kcal/801kJ; **Protein** 2.5g; **Carbohydrate** 22.6g, of which sugars 8.3g; **Fat** 10.8g, of which saturates 1.4g; **Cholesterol** 16mg; **Calcium** 37mg; **Fibre** 0.6g; **Sodium** 11mg.

Deep-fried cherries

Crispy on the outside, soft in the middle, these cherries make an unusual dessert.

INGREDIENTS

Serves 4–6

450g/1lb ripe red cherries, on their stalks

120g/4¹/₄oz/generous 1 cup plain (all-purpose) flour

60g/2¹/₄oz/generous ¹/₄ cup golden caster (superfine) sugar

75ml/5 tbsp full-fat (whole) milk

75ml/5 tbsp dry white wine

3 eggs, beaten

vegetable oil, for deep-frying

icing (confectioners') sugar and ground cinnamon, for dusting

1 Wash and dry the cherries. Tie the stalks together into clusters of 4–5 cherries. Place the flour, golden caster sugar, milk, white wine and eggs in a bowl and mix to make a smooth batter. Pour the vegetable oil into a deep-fat fryer or pan and heat to 190°C/375°F.

2 Working in batches, half-dip each cherry cluster into the batter and then carefully drop the cluster into the hot oil. Fry for 3–4 minutes until golden. Remove the cherries with a slotted spoon and drain on a wire rack placed over kitchen paper.

3 Dust the cherries with icing sugar and cinnamon and serve with ice cream.

Energy 201kcal/840kJ; **Protein** 3.7g; **Carbohydrate** 25.7g, of which sugars 7.3g; **Fat** 10g, of which saturates 1.3g; **Cholesterol** 26mg; **Calcium** 46mg; **Fibre** 1.3g; **Sodium** 11mg.

OILS IN STIR-FRIED DISHES

Oriental cuisine is so healthy because it involves frying ingredients in small amounts of hot oil to preserve the nutrients and keep the final dish light and crisp.

Duck and sesame stir-fry

This stir-fry is cooked using a mixture of half-and-half sesame and vegetable oil.

INGREDIENTS

Serves 4

15ml/1 tbsp sesame oil

15ml/1 tbsp vegetable oil

4 garlic cloves, finely sliced

250g/9oz duck meat, cut into bitesize pieces

2.5ml/½ tsp dried chilli flakes

15ml/1 tbsp Thai fish sauce (nam pla)

15ml/1 tbsp light soy sauce

120ml/4fl oz/½ cup water

1 head broccoli, cut into small florets

coriander (cilantro) and 15ml/1 tbsp toasted
   sesame seeds, to garnish

1 Heat the oils in a wok and gently stir-fry the garlic until it is golden brown. Add the duck and stir-fry for a further 2 minutes, until it begins to brown.

2 Stir in the chilli flakes, fish sauce, soy sauce and water. Add the broccoli and continue to stir-fry for 2 minutes, until the duck is cooked through. Serve garnished with coriander and sesame seeds.

Energy 192kcal/798kJ; **Protein** 18.7g;
Carbohydrate 2.7g, of which sugars 2.3g; **Fat**
12.9g, of which saturates 2.1g; **Cholesterol** 69mg;
Calcium 104mg; **Fibre** 3.6g; **Sodium** 436mg.

Celebration noodles

Stir-fry the ingredients in this noodle dish in palm or coconut oil for an authentic Filipino taste.

INGREDIENTS

Serves 4

30ml/2 tbsp palm or coconut oil

1 large onion, finely chopped

2–3 garlic cloves, finely chopped

250g/9oz pork loin, cut into thin strips

250g/9oz fresh shelled shrimps

2 carrots, cut into matchsticks

½ small green cabbage, finely shredded

250ml/8fl oz/ 1 cup pork or chicken stock

50ml/2fl oz/¼ cup soy sauce

15ml/1 tbsp palm sugar (jaggery)

450g/1lb fresh egg noodles

2 hard-boiled eggs, finely chopped

1 lime, quartered

1 Heat 15ml/1 tbsp oil in a wok, stir in the onion and garlic and fry until fragrant and beginning to colour.

2 Toss in the pork and shrimp and stir-fry for 2 minutes, then transfer the mixture on to a plate.

3 Return the wok to the heat, add the remaining oil, then stir in the carrots and cabbage and stir-fry for 2–3 minutes. Transfer the vegetables on to the plate with the pork and shrimp.

4 Pour the stock, soy sauce and sugar into the wok and stir until the sugar has dissolved. Add the noodles and cook for about 3 minutes, until tender but still firm to the bite. Toss in the pork, shrimp, cabbage and carrots, mixing thoroughly.

5 Transfer the noodles on to a serving dish and scatter the chopped eggs over the top. Serve immediately with the lime wedges.

Energy 728kcal/3069kJ; **Protein** 43.6g;
Carbohydrate 98g, of which sugars 16.9g; **Fat**
20.8g, of which saturates 5g; **Cholesterol** 290mg;
Calcium 159mg; **Fibre** 6.3g; **Sodium** 1303mg.

Thai fried rice

This dish uses two different flavoured oils. The ingredients are first stir-fried in groundnut oil, while a small amount of hot and spicy chilli oil is added as a last minute condiment.

INGREDIENTS

Serves 4

475ml/16fl oz/2 cups water

50g/2oz/¹/₂ cup coconut milk powder

350g/12oz/1³/₄ cups Thai jasmine
   rice, rinsed

30ml/2 tbsp groundnut (peanut) oil

2 garlic cloves, chopped

1 small onion, finely chopped

2.5cm/1in piece fresh root ginger, peeled
   and grated

225g/8oz skinned chicken breast fillets, cut
   into 1cm/¹/₂in pieces

1 red (bell) pepper, seeded and sliced

115g/4oz/1 cup drained canned whole
   kernel corn

5ml/1 tsp chilli oil

5ml/1 tsp hot curry powder

2 eggs, beaten

salt

spring onion (scallion) shreds, to garnish

1 Pour the water into a pan and whisk in the coconut milk powder. Add the rice and bring to the boil. Reduce the heat, cover and cook for 12 minutes, or until the rice is tender and the liquid has been absorbed. Spread the rice on a baking sheet and leave until cold.

2 Heat the groundnut oil in a wok, add the garlic, onion and ginger and stir-fry over a medium heat for 2 minutes.

3 Push the onion mixture to the sides of the wok, add the chicken to the centre and stir-fry for 2 minutes. Add the rice and toss well. Stir-fry over a high heat for about 3 minutes more, until the chicken is cooked through.

4 Stir in the sliced red pepper, corn, chilli oil and curry powder, with salt to taste. Toss over the heat for 1 minute. Stir in the beaten eggs and cook for 1 minute more. Garnish with the spring onion shreds and serve.

Energy 669kcal/2,775kJ; **Protein** 48.7g; **Carbohydrate** 5g, of which sugars 3.8g; **Fat** 45.9g, of which saturates 11.1g; **Cholesterol** 250mg; **Calcium** 37mg; **Fibre** 0.9g; **Sodium** 196mg.

Stir-fried pork with dried shrimp

Classic Chinese ingredients are stir-fried in vegetable oil.

INGREDIENTS

Serves 4

250g/9oz pork fillet (tenderloin), sliced

30ml/2 tbsp vegetable oil

2 garlic cloves, finely chopped

45ml/3 tbsp dried shrimp

10ml/2 tsp dried shrimp paste

30ml/2 tbsp soy sauce

juice of 1 lime

15ml/1 tbsp light muscovado (brown) sugar

1 small fresh red or green chilli, seeded and
   finely chopped

4 pak choi (bok choy), shredded

1 Place the pork in the freezer for 30 minutes, until firm. Cut it into thin slices.

2 Heat the oil in a wok or frying pan and cook the garlic until golden brown. Add the pork and stir-fry for about 4 minutes, until just cooked through.

3 Add the dried shrimp, then stir in the shrimp paste, with the soy sauce, lime juice and sugar.

4 Add the chilli and pak choi and toss over the heat until the vegetables are just wilted. Transfer to individual bowls and serve immediately.

Energy 200kcal/833kJ; **Protein** 23.1g; **Carbohydrate** 6.3g, of which sugars 6.2g; **Fat** 9.2g, of which saturates 1.7g; **Cholesterol** 96mg; **Calcium** 334mg; **Fibre** 2.4g; **Sodium** 1,223mg.

OILS IN SHALLOW-FRIED DISHES

Cooking quickly, slowly or sautéing, shallow-frying is a very versatile kitchen technique which is used in many favourite dishes from all around the world.

Meatballs in rich tomato sauce

This dish is an Italian favourite. Simmering the meatballs in a rich, oil-based sauce keeps them mouth-wateringly tender.

INGREDIENTS

Serves 4

500g/1¼lb minced (ground) beef or lamb

1 onion, grated

1 egg, lightly beaten

50g/2oz/generous ⅓ cup short
   grain rice

45ml/3 tbsp chopped flat leaf parsley

finely grated rind of ½ orange, plus extra
   to garnish (optional)

salt and ground black pepper

For the sauce

60ml/4 tbsp extra virgin olive oil

1 onion, thinly sliced

3–4 fresh sage leaves, finely sliced

400g/14oz can tomatoes

300ml/½ pint/1¼ cups beef stock

1 Put the meat in a bowl and add the onion, egg, rice and parsley. Add the grated orange rind with the salt and pepper. Mix all the ingredients well, then shape the mixture into round balls.

2 Make the sauce. Heat the oil in a wide pan that will take the meatballs in one layer. Sauté the onion slices until they are golden. Add the sage, then the tomatoes, breaking them up with a wooden spoon.

3 Simmer for a few minutes, then add the stock and bring to the boil. Lower the meatballs into the sauce. Do not stir but rotate the pan to coat evenly. Season, cover the pan and simmer for 30 minutes until the sauce has thickened. Scatter over a little orange rind to garnish, if you like.

Energy 475kcal/1972kJ; **Protein** 28.5g; **Carbohydrate** 15.8g, of which sugars 5.1g; **Fat** 33.2g, of which saturates 10.7g; **Cholesterol** 123mg; **Calcium** 58mg; **Fibre** 2g; **Sodium** 131mg.

Asian-style crab cakes

Shallow-frying these crab cakes gives them a lovely crispy finish.

INGREDIENTS

Makes 16

450g/1lb/2⅔ cups fresh crab meat

15ml/1 tbsp grated fresh root ginger

15–30ml/1–2 tbsp plain (all-purpose) flour

60ml/4 tbsp groundnut (peanut) or
   sunflower oil

salt and ground black pepper

1 Put the crab meat in a bowl with the ginger, salt and ground black pepper and flour. Stir well until mixed. Using floured hands, divide into 16 equal-sized pieces and shape roughly into patties.

2 Heat the oil in a frying pan and add the patties, four at a time. Cook for 3 minutes on each side, until golden. Remove with a metal spatula and leave to drain on kitchen paper for a few minutes.

3 Keep warm while you cook the remaining patties in the same way. Serve.

Energy 67kcal/280kJ; **Protein** 5.7g; **Carbohydrate** 1.5g, of which sugars 0g; **Fat** 4.3g, of which saturates 0.5g; **Cholesterol** 20mg; **Calcium** 3mg; **Fibre** 0.1g; **Sodium** 119mg.

Flash-fried squid

Squid is first marinaded then fried
very quickly in olive oil.

INGREDIENTS

Serves 6–8

500g/1¼lb very small squid, cleaned

90ml/6 tbsp olive oil, plus extra

1 red chilli, seeded and finely chopped

10ml/2 tsp Spanish smoked paprika (pimentón)

30ml/2 tbsp plain (all-purpose) flour

2 garlic cloves, finely chopped

15ml/1 tbsp sherry vinegar

5ml/1 tsp grated lemon rind

30–45ml/2–3 tbsp finely chopped fresh parsley

1 Cut the squid body sacs into rings and
the tentacles into bitesize pieces. Place in
a bowl and add 30ml/2 tbsp of the oil,
half the chilli and the paprika. Season,
cover and marinate for 2 hours.

2 Toss the squid in the flour and divide
it into 2 batches. Heat the remaining
oil in a wok. Add the first batch
of squid and fry for 2 minutes, until
the squid becomes opaque and the
tentacles have curled. Sprinkle in half
the garlic. Stir, then turn out on to a plate.

3 Repeat with the second batch.
Sprinkle over the sherry vinegar, lemon
rind, remaining chilli and parsley. Serve.

Energy 139kcal/580kJ; **Protein** 10.1g;
Carbohydrate 3.8g, of which sugars 0.1g; **Fat** 9.4g,
of which saturates 1.4g; **Cholesterol** 141mg;
Calcium 21mg; **Fibre** 0.3g; **Sodium** 70mg.

Griddled beef with sesame

In this Korean dish, thin strips of sirloin
are marinated in sesame oil and soy
sauce, then cooked in a griddle pan.

INGREDIENTS

Serves 4

800g/1¾lb sirloin steak

For the marinade

4 spring onions (scallions)

½ onion

1 Asian pear

60ml/4 tbsp dark soy sauce

60ml/4 tbsp sugar

30ml/2 tbsp sesame oil

10ml/2 tsp ground black pepper

5ml/1 tsp sesame seeds

2 garlic cloves, crushed

15ml/1 tbsp lemonade

1 Finely slice the steak, and tenderize
by bashing with a meat mallet. Cut
into bitesize strips.

2 Shred one of the spring onions and
set aside for a garnish. Finely slice
the remaining spring onions, the onion
and pear.

3 Combine all the marinade ingredients
in a large bowl to form a paste, adding
a little water if necessary.

4 Mix the beef in with the marinade,
making sure that it is well coated. Leave
in the refrigerator for at least 30
minutes or up to 2 hours (if left longer
the meat will become too salty).

5 Heat a griddle pan gently. Add the
meat and cook over a medium heat.
Once the meat is cooked through,
transfer it to a large serving dish, garnish
with the spring onion and serve.

Energy 330kcal/1382kJ; **Protein** 47.3g;
Carbohydrate 8.2g, of which sugars 8.1g; **Fat** 12.1g,
of which saturates 4.5g; **Cholesterol** 102mg;
Calcium 22mg; **Fibre** 0.2g; **Sodium** 141mg.

OILS IN GRILLED DISHES

Use oil for marinading meat or fish before cooking, or brush with oil before grilling to stop food drying out. Basting with oil during cooking will keep meat tender.

Monkfish kebabs

In this recipe, monkfish is marinated in a mixture of olive oil and lemon juice.

INGREDIENTS

Serves 4

900g/2lb fresh monkfish tail, skinned

3 (bell) peppers, red, green and yellow

juice of 1 lemon

60ml/4 tbsp olive oil

salt and ground black pepper

1 Trim the skinned monkfish and cut it into bitesize cubes. Cut each pepper into quarters, and then seed and halve each quarter. Combine the lemon juice and oil in a bowl and add seasoning.

2 Marinate the fish and pepper in the mixture for 20 minutes. Soak four wooden skewers in cold water for 30 minutes to prevent them burning during cooking.

3 Preheat a very hot grill (broiler). Thread pieces of fish and pepper alternately. Cook for 10 minutes, turning and basting frequently. Serve in pitta bread, with a squeeze of fresh lemon juice.

Energy 377kcal/1,586kJ; **Protein** 37.4g; **Carbohydrate** 24.1g, of which sugars 23.1g; **Fat** 15.3g, of which saturates 2.4g; **Cholesterol** 32mg; **Calcium** 54mg; **Fibre** 2.8g; **Sodium** 382mg.

Grilled hake with lemon and chilli

The oil in this dish infuses the fish with the flavours of chilli and lemon rind.

INGREDIENTS

Serves 4

4 hake fillets, each 150g/5oz

30ml/2 tbsp olive oil

finely grated rind and juice
  of 1 unwaxed lemon

15ml/1 tbsp crushed chilli flakes

salt and ground black pepper

1 Preheat the grill (broiler) to high. Brush the hake fillets all over with the olive oil and place them skin side up on a baking sheet.

2 Grill (broil) the fish for 4–5 minutes, until the skin is crispy, then carefully turn them over using a metal spatula.

3 Sprinkle the fillets with the lemon rind and chilli flakes and season with salt and ground black pepper.

4 Grill the fillets for a further 2–3 minutes, or until the hake is cooked through. (Test using the point of a sharp knife; the flesh should flake.) Squeeze over the lemon juice just before serving.

Energy 206kcal/862kJ; **Protein** 27g; **Carbohydrate** 0g, of which sugars 0g; **Fat** 11g, of which saturates 2g; **Cholesterol** 35mg; **Calcium** 26mg; **Fibre** 0g; **Sodium** 300mg.

Fragrant grilled chicken

The sesame oil marinade adds flavour to this delicately-spiced chicken. Basting the chicken with the marinade mixture halfway through cooking will ensure that the chicken remains tender and full of flavour.

INGREDIENTS

Serves 4

450g/1lb boneless chicken breast portions, with the skin on

30ml/2 tbsp sesame oil

2 garlic cloves, crushed

2 coriander (cilantro) roots, finely chopped

2 small fresh red chillies, seeded and finely chopped

30ml/2 tbsp Thai fish sauce (nam pla)

5ml/1 tsp sugar

cooked rice, to serve

lime wedges, to garnish

For the sauce

90ml/6 tbsp rice vinegar

60ml/4 tbsp sugar

2.5ml/½ tsp salt

2 garlic cloves, crushed

1 small fresh red chilli, seeded and finely chopped

115g/4oz/4 cups fresh coriander (cilantro), finely chopped

1 Lay the chicken between two sheets of clear film (plastic wrap) and beat with the side of a rolling pin until the meat is about half its original thickness. Place in a large, shallow dish or bowl.

2 Mix together the sesame oil, garlic, coriander roots, red chillies, fish sauce and sugar in a jug (pitcher), stirring until the sugar has dissolved. Pour over the chicken and turn to coat. Cover with clear film and marinate for at least 20 minutes. Meanwhile, make the sauce.

3 Heat the vinegar in a small pan, add the sugar and stir until dissolved. Add the salt and stir until the mixture begins to thicken. Add the remaining sauce ingredients, stir well, then spoon the sauce into a serving bowl.

4 Preheat the grill (broiler) and cook the chicken for 5 minutes. Turn and baste with the marinade, then cook for 5 minutes more, or until cooked through and golden. Serve with rice and the sauce, garnished with lime wedges.

Energy 243kcal/1,022kJ; **Protein** 28g; **Carbohydrate** 17.7g, of which sugars 17.6g; **Fat** 7.1g, of which saturates 1.2g; **Cholesterol** 79mg; **Calcium** 73mg; **Fibre** 1.5g; **Sodium** 502mg.

Rosemary scented lamb

The lamb in this recipe is marinaded overnight in an oil-based mixture.

INGREDIENTS

Serves 4–8

2 x 8-chop racks of lamb, French trimmed

8 large fresh rosemary sprigs

2 garlic cloves, finely sliced

90ml/6 tbsp extra virgin olive oil

30ml/2 tbsp red wine

salt and ground pepper

1 Cut the racks into eight portions, each consisting of two linked chops, and tie a rosemary sprig to each one. Lay them in a single layer in a bowl.

2 Mix the garlic, oil and wine, and pour over the lamb. Cover and chill overnight, turning as often as possible.

3 Bring the lamb to room temperature 1 hour before cooking. Remove the lamb from the marinade. Season the meat 15 minutes before cooking.

4 Grill (broil) the chops for 5 minutes on each side. Remove from the grill (broiler), cover and rest for 10 minutes. Serve.

Energy 433kcal/1788kJ; **Protein** 23.4g; **Carbohydrate** 0g, of which sugars 0g; **Fat** 37.6g, of which saturates 16.4g; **Cholesterol** 101mg; **Calcium** 17mg; **Fibre** 0g; **Sodium** 83mg.

OILS IN ROASTED DISHES

Adding oil to roasted dishes will stop them becoming too dry in the oven. It can be used with other ingredients, or on its own to give a nice crisp finish to vegetables or meat.

Roast potatoes

A classic accompaniment to roast meat, roast potatoes are coated thoroughly in olive oil to ensure a crisp finish.

INGREDIENTS

Serves 4

1.3kg/3lb floury potatoes

90ml/6 tbsp olive oil

salt

1 Preheat the oven to 200°C/400°F/ Gas 6. Peel the potatoes and cut them into chunks.

2 Boil the potatoes in salted water for about 5 minutes, drain, return to the pan, and shake them to roughen the surfaces.

3 Put the oil into a large roasting pan and put into the hot oven to heat. Add the potatoes, coating them in the oil.

4 Return to the oven and cook for 40–50 minutes, turning once or twice, until crisp and cooked through.

Energy 484kcal/2,048kJ; **Protein** 9.4g; **Carbohydrate** 84.2g, of which sugars 2g; **Fat** 14.6g, of which saturates 5.9g; **Cholesterol** 13mg; **Calcium** 26mg; **Fibre** 5.9g; **Sodium** 29mg.

Sardines with tomatoes, thyme and purple basil

A mix of tomatoes, spices and oil infuses the fish with flavour as they roast.

INGREDIENTS

Serves 4

8 large sardines, scaled, gutted and
  thoroughly washed

6–8 fresh thyme sprigs

juice of ¹/₂ lemon

2 x 400g/14oz cans chopped tomatoes,
  drained of juice

60–75ml/4–5 tbsp olive oil

4 garlic cloves, smashed flat

5ml/1 tsp sugar

1 bunch of fresh purple basil

salt and ground black pepper

lemon wedges, to serve

1 Preheat the oven to 180°C/350°F/ Gas 4. Lay the sardines side by side in an ovenproof dish, place a sprig of thyme between each one and squeeze the lemon juice over them.

2 In a large bowl, mix together the chopped tomatoes, olive oil, garlic and sugar. Season and stir in most of the basil leaves, then pour the mixture over the sardines.

3 Bake, uncovered, for 25 minutes. Sprinkle the remaining basil leaves over the top and serve hot, with the lemon wedges.

Energy 219kcal/915kJ; **Protein** 11.7g; **Carbohydrate** 7.3g, of which sugars 7.3g; **Fat** 16.2g, of which saturates 3.1g; **Cholesterol** 0mg; **Calcium** 57mg; **Fibre** 2g; **Sodium** 78mg.

Aubergines in a chilli sauce

The aubergines are brushed with coconut oil before roasting to keep them moist.

INGREDIENTS

Serves 4

4 small aubergines (eggplants), butterflied

60ml/4 tbsp coconut oil

4 shallots, finely chopped

4 garlic cloves, finely chopped

25g/1oz fresh root ginger, finely chopped

3–4 red chillies, seeded and finely chopped

400g/14oz can tomatoes, drained

5–10ml/1–2 tsp palm sugar (jaggery)

juice of 2 limes

salt

1 small bunch fresh coriander (cilantro), finely chopped, to garnish

1 Preheat the oven to 180°C/350°F/ Gas 4. Put the aubergines on a baking tray, brush with 30ml/2 tbsp of the coconut oil and bake for 40 minutes, until soft.

2 Using a mortar and pestle, grind the shallots, garlic, ginger and chillies to a paste. Heat the remaining 30ml/2 tbsp of oil in a wok and stir in the spice paste and cook for 1–2 minutes. Add the tomatoes and sugar and cook for 3–4 minutes, then stir in the lime juice and salt to taste.

3 Put the aubergines on to a serving dish and spoon the sauce over them. Garnish with the chopped coriander and serve.

Energy 100kcal/419kJ; **Protein** 2,1g;
Carbohydrate 9.4g, of which sugars 8.8g;
Fat 6.4g, of which saturates 0.9g; **Cholesterol** 0mg;
Calcium 42mg; **Fibre** 3.7g; **Sodium** 15mg.

Roast shoulder of lamb with whole garlic cloves

Olive oil is drizzled over both potatoes and lamb in this recipe.

INGREDIENTS

serves 4–6

675g/1½lb waxy potatoes, peeled and cut into large dice

12 garlic cloves, unpeeled

45ml/3 tbsp olive oil

1 whole shoulder of lamb

salt and ground black pepper

1 Preheat the oven to180°C/350°F/ Gas 4. Put the potatoes and garlic cloves into a large roasting pan and season with salt and pepper.

2 Pour over 30ml/2 tbsp of the oil and toss the potatoes and garlic to coat.

3 Place a rack over the roasting pan, so that it is not touching the potatoes at all. Carefully place the shoulder of lamb on the rack and drizzle over the remaining oil. Season with salt and pepper.

4 Roast the lamb and potatoes in the oven for about 2–2½ hours, or until the lamb is cooked through.

5 Halfway through the cooking time, carefully take the lamb and the rack off the roasting pan and turn the potatoes to ensure even cooking, replacing the lamb and the rack after turning them.

Energy 668kcal/2775kJ; **Protein** 29.2g;
Carbohydrate 20.8g, of which sugars 1.7g; **Fat** 52.6g, of which saturates 24.1g; **Cholesterol** 113mg; **Calcium** 22mg; **Fibre** 1.8g; **Sodium** 123mg.

OILS IN CAKES AND BISCUITS

In these recipes, oil is used to make a dough or batter and imparts wonderful rich flavours and added moistness to a range of sweet baking mixtures.

Olive oil biscuits

These biscuits, made with an oil-based dough, are a Portuguese favourite.

INGREDIENTS

Makes about 20

6 eggs

150ml/¼ pint/²⁄₃ cup olive oil

100g/3¾oz/generous ½ cup sugar

15ml/1 tbsp brandy

about 250g/9oz/2¼ cups plain (all-purpose)
  flour, plus extra for dusting

1 Preheat the oven to 180°C/350°F/
Gas 4. Beat the eggs with the olive oil,
the sugar and brandy with an electric
mixer until smooth.

2 Gradually beat in the flour on a low
speed until a dough forms.

3 Fill moulds and bake for 20 to 30
minutes (depending on the size of the
moulds), until golden.

4 Leave the biscuits to stand for
2 minutes and remove from the moulds.

Energy 111kcal/465kJ; **Protein** 3g;
Carbohydrate 9.7, of which sugars 0.2g; **Fat** 6.8g,
of which saturates 1.2g; **Cholesterol** 57.1mg;
Calcium 26mg; **Fibre** 0.4g; **Sodium** 21.4mg.

Apple cake

This traditional Polish cake is firm and
moist, with pieces of apple peeking
through the top.

INGREDIENTS

Serves 6–8

375g/13oz/3¼ cups self-raising
  (self-rising) flour

3–4 large cooking apples,

10ml/2 tsp ground cinnamon

500g/1¼lb/2½ cups caster
  (superfine) sugar

4 eggs, lightly beaten

250ml/8fl oz/1 cup vegetable oil

120ml/4fl oz/½ cup orange juice

10ml/2 tsp vanilla extract

2.5ml/½ tsp salt

1 Preheat the oven to 180°C/350°F/
Gas 4. Grease a 30–8cm/12–15in square
cake tin (pan) and dust with a little of
the flour.

2 Core and thinly slice the apples,
but do not peel. Put the sliced
apples in a bowl and mix with the
cinnamon and 75ml/5 tbsp of the sugar.

3 In a separate bowl, beat together the
eggs, remaining sugar, vegetable oil,
orange juice and vanilla extract until
well combined. Sift in the remaining
flour and salt, then stir into the mixture.

4 Pour two-thirds of the cake mixture
into the prepared tin, top with one-third
of the apples, then pour over the
remaining cake mixture and top with
the remaining apple. Bake for 1 hour, or
until golden brown.

5 Cool slightly in the tin to allow the
juices to soak in. Serve cut into squares.

Energy 653kcal/2751kJ; **Protein** 7.8g;
Carbohydrate 105.4g, of which sugars 70.6g; **Fat**
25.3g, of which saturates 3.4g; **Cholesterol** 95mg;
Calcium 215mg; **Fibre** 2.1g; **Sodium** 210mg.

Semolina cake

This is a traditional Greek cake recipe. It takes very little time to make and uses inexpensive ingredients that most Greek kitchens would have in stock.

INGREDIENTS

Serves 6–8

500g/1¼lb/2¾ cups caster (superfine) sugar

1 litre/1¾ pints/4 cups cold water

1 cinnamon stick

250ml/8fl oz/1 cup olive oil

350g/12oz/2 cups coarse semolina

50g/2oz/½ cup blanched almonds

30ml/2 tbsp pine nuts

5ml/1 tsp ground cinnamon

1 Put the sugar in a heavy pan, pour in the water and add the cinnamon stick. Bring to the boil, stirring until the sugar dissolves, then boil without stirring for about 4 minutes to make a syrup.

2 Meanwhile, heat the oil in a separate, heavy pan. When it is almost smoking, add the semolina gradually and stir constantly until it turns light brown.

3 Lower the heat, add the almonds and pine nuts, and brown together for 2–3 minutes, stirring constantly.

4 Take the semolina mixture off the heat and set aside. Remove the cinnamon stick from the hot syrup using a slotted spoon and discard it.

5 Protecting your hand with an oven glove or dish towel, carefully add the hot syrup to the semolina mixture a little at a time, stirring constantly. The mixture will probably hiss and spit at this point, so stand well away.

6 Return the pan to a gentle heat and stir until all the syrup has been absorbed and the mixture looks nice and smooth.

7 Remove the pan from the heat, cover it with a clean dish towel and let it stand for 10 minutes so that any remaining moisture is absorbed.

8 Scrape the mixture into a 20–23cm/ 8–9in round cake tin (pan), preferably fluted, and set it aside.

9 When it is cold, unmould it on to a serving platter and dust it evenly all over with the ground cinnamon. Serve in slices.

Energy 888kcal/3,731kJ; **Protein** 9.1g; **Carbohydrate** 133.1g, of which sugars 87.6g; **Fat** 39.1g, of which saturates 4.9g; **Cholesterol** 0mg; **Calcium** 75mg; **Fibre** 1.9g; **Sodium** 13mg.

Spiced apple cake

Grated apple and chopped dates give this cake a natural sweetness.

INGREDIENTS

Serves 8

225g/8oz/2 cups self-raising (self-rising) wholemeal (wholewheat) flour

5ml/1 tsp baking powder (baking soda)

10ml/2 tsp ground cinnamon

175g/6oz/1 cup chopped dates

75g/3oz/½ cup light muscovado (brown) sugar

15ml/1 tbsp pear and apple spread

120ml/4fl oz/½ cup apple juice

2 eggs

90ml/6 tbsp sunflower oil

2 eating apples, cored and grated

15ml/1 tbsp chopped walnuts

1 Preheat the oven to 180°C/350°F/Gas 4. Grease and line a 20cm/8in cake tin (pan). Sift the flour, baking powder and cinnamon into a bowl, mix in the dates and make a well in the centre.

2 Mix the sugar with the pear and apple spread in a small bowl. Gradually stir in the apple juice. Add to the dry ingredients with the eggs, oil and apples. Mix.

3 Spoon the mixture into the prepared tin, sprinkle with walnuts and bake for 1 hour, or until a skewer inserted into the cake comes out clean. Cool on a wire rack.

Energy 282kcal/1186kJ; **Protein** 4.9g; **Carbohydrate** 42.8g, of which sugars 21.3g; **Fat** 11.3g, of which saturates 1.5g; **Cholesterol** 48mg; **Calcium** 60mg; **Fibre** 1.6g; **Sodium** 22mg.

OILS IN BREADS

Olive oil is an essential ingredient in the breads of the Mediterranean region, particularly those of Greece and Italy. The addition of oil brings both moisture and flavour.

Italian olive oil bread

This classic Italian bread, known as Pugliese, is moistened and flavoured with fruity extra virgin olive oil.

INGREDIENTS

Makes one large loaf

For the biga starter

175g/6oz/1½ cups unbleached white
    bread flour

7g/¼ oz fresh yeast

90ml/6 tbsp lukewarm water

For the dough

225g/8oz/2 cups unbleached white bread
    flour, plus extra for dusting

225g/8oz/2 cups unbleached wholemeal
    (whole-wheat) bread flour

5ml/1 tsp caster (superfine) sugar

10ml/2 tsp salt

15g/½ oz fresh yeast

275ml/9fl oz/generous 1 cup lukewarm water

75ml/5 tbsp extra virgin olive oil

1 Sift the flour for the biga starter into a large bowl. Make a well in the centre. In a small bowl, cream the yeast with the water. Pour the liquid into the centre of the flour and gradually mix in the surrounding flour to form a firm dough.

2 Turn the dough out on to a lightly floured surface and knead for 5 minutes until smooth and elastic. Return to the bowl, cover with lightly oiled clear film (plastic wrap) and leave to rise, in a warm place, for 8–10 hours, or until the dough has risen well and is starting to collapse.

3 Lightly flour a baking sheet. Mix the flours, sugar and salt for the dough in a large bowl. Cream the yeast and the water in another large bowl, then stir in the biga and mix together.

4 Stir in the flour mixture a little at a time, then add the olive oil in the same way, and mix to a soft dough. Turn out on to a lightly floured surface and knead the dough for 8–10 minutes until smooth and elastic.

5 Place in a lightly oiled bowl, cover with lightly oiled clear film and leave to rise, in a warm place, for 1–1½ hours, or until doubled in bulk.

6 Turn out on to a lightly floured surface and knock back (punch down). Gently pull out the edges and fold under to make a round. Transfer to the prepared baking sheet, cover with lightly oiled clear film and leave to rise, in a warm place, for 1–1½ hours, or until almost doubled in size.

7 Meanwhile, preheat the oven to 230°C/450°F/Gas 8. Lightly dust the loaf with flour and bake for 15 minutes. Reduce the oven temperature to 200°C/400°F/Gas 6 and bake for a further 20 minutes, or until the loaf sounds hollow when tapped on the base. Transfer to a wire rack to cool.

Energy 2502kcal/10567kJ; **Protein** 72g; **Carbohydrate** 430.4g, of which sugars 11.8g; **Fat** 66.7g, of which saturates 9.5g; **Cholesterol** 0mg; **Calcium** 468mg; **Fibre** 43g; **Sodium** 3949mg.

Panini all'olio

The name of these little rolls literally translates as 'small oil loaf'.

INGREDIENTS

Makes 16 rolls

60ml/4 tbsp extra virgin olive oil

450g/1lb/4 cups unbleached strong white
  bread flour

10ml/2 tsp salt

15g/¹/₂ oz fresh yeast

1 Lightly oil three baking sheets. Sift the flour and salt together in a bowl and make a well in the centre. Measure 250ml/8fl oz/1 cup lukewarm water.

2 Cream the yeast with half the water, then stir in the remainder. Add to the well with the oil and mix to a dough.

3 Turn the dough out on to a lightly floured surface and knead for 10 minutes, until smooth and elastic. Place in a lightly oiled bowl, cover with lightly oiled clear film (plastic wrap) and leave to rise for 1 hour, or until doubled in bulk.

4 Turn on to a lightly floured surface and knock back (punch down). Divide into 12 equal pieces and shape into rolls.

5 To make twists, roll each piece of dough into a strip 30cm/12in long and 4cm/1¹/₂in wide. Twist into a spiral and join the ends to make a circle. Place on the baking sheets, spaced well apart. Brush lightly with olive oil, cover with lightly oiled clear film and leave to rise for 30 minutes.

6 To make fingers, flatten each piece of dough into an oval and roll to 23cm/9in long. Roll up from the wider end. Gently stretch the dough roll to 23cm/9in long. Cut in half. Place on the baking sheets, spaced well apart. Brush the dough with olive oil, cover with lightly oiled clear film and leave to rise for 30 minutes.

7 To make artichoke-shapes, shape each piece of dough into a ball and space well apart on the baking sheets. Brush with oil, cover with lightly oiled clear film and leave to rise 30 minutes. Using scissors, snip 5mm/¹/₄ in deep cuts in a circle on the top of each ball, then make five larger horizontal cuts around the sides.

8 Preheat the oven to 200°C/400°F/Gas 6. Bake the rolls for 15 minutes.

Energy 121kcal/509kJ; **Protein** 2.6g; **Carbohydrate** 21.9g, of which sugars 0.4g; **Fat** 3.1g, of which saturates 0.5g; **Cholesterol** 0mg; **Calcium** 39mg; **Fibre** 0.9g; **Sodium** 246mg.

Polenta and pepper bread

This bread is delicious when drizzled with a little olive oil.

INGREDIENTS

Makes 2 loaves

175g/6oz/1¹/₂ cups polenta

5ml/1 tsp salt

350g/12oz/3 cups unbleached strong plain
  flour, plus extra for dusting

5ml/1 tsp sugar

7g/¹/₄oz sachet easy-blend dried yeast

1 red (bell) pepper, roasted, peeled and diced

15ml/1 tbsp olive oil

1 Mix the polenta, salt, flour, sugar and yeast in a bowl. Stir in the pepper until it is evenly distributed, then make a well in the centre of the mixture. Grease two loaf tins.

2 Add 300ml/¹/₂ pint/1¹/₄ cups warm water and the olive oil and mix to a soft dough. Knead for 10 minutes until smooth and elastic. Place in an oiled bowl, cover with oiled clear film (plastic wrap) and leave to rise in a warm place for 1 hour until doubled. Knock back the dough, knead lightly, then divide in two.

3 Shape each piece into an oblong and place in the tins. Cover with oiled clear film and leave to rise for 45 minutes. Preheat the oven to 220°C/425°F/Gas 7. Bake for 30 minutes until golden. Leave for 5 minutes, then cool on a wire rack.

Energy 1994kcal/8427kJ; **Protein** 50.5g; **Carbohydrate** 366.9g, of which sugars 7.1g; **Fat** 44.7g, of which saturates 15.1g; **Cholesterol** 53mg; **Calcium** 1085mg; **Fibre** 14.9g; **Sodium** 2337mg.

INDEX

ACKNOWLEDGMENTS & PICTURE CREDITS

The author and publishers would like
to thank the following companies for
supplying vinegars and oils:
Aspall, Belazu, Blue Dragon, Clearspring, Delicious Fine Foods, Fieldhouse Farms, Hillfarm
Oils, Meridian Foods, Suzanne's Vinegars.
Models: Kiera Blakey and Patrick Tubbritt.

The publisher would like to thank the
following picture libraries for the use of
their pictures in the book.
l=left, r=right, t=top, b=bottom, bl=bottom
left, br=bottom right, bm=bottom middle,
m=middle, tm=top middle, tl=top left,
tr=top right.

Alamy: 12b, 18, 21l, 22t, 28l, 28m,
29l, 39br, 44tr, 73bl, 88tr, 88br, 139tr,
152bl, 163tl.

The Art Archive: 14t.

Bridgeman Art Library: 12t, 13t, 13b,
20, 134b, 135bl, 136b, 137l, 138bl, 144t.

Corbis: 14b, 17b, 21m, 31l, 31m, 138br,
142tr, 152br, 160tr, 179tl.

Getty Images: 36m, 37tl, 39bl, 43t.

Istockphoto: 76tr, 78br, 81tl, 81tr, 85tm,
85bm, 134t, 135br, 139tl, 143b, 144b,
145b, 151tl, 151tm, 151tr, 152bm, 154b,
156bl, 157tl, 158tl, 159tl, 163bl, 163br,
204br, 205tm, 205bm, 206br, 207br,
210t, 219tl.